GROW
FRUITS & VEGETABLES THE WAY THEY USED TO TASTE

by John F. Adams

AN ESSAY ON BREWING
BEEKEEPING—THE GENTLE CRAFT
BACKYARD POULTRY RAISING
TWO PLUS TWO EQUALS MINUS SEVEN

GROW FRUITS & VEGETABLES THE WAY THEY USED TO TASTE

John F. Adams

WYNWOOD™ Press
New York, New York

This book was originally published under the title *Guerrilla Gardening.*

Library of Congress Cataloging-in-Publication Data

Adams, John Festus.
Grow fruits and vegetables the way they used to taste / John F. Adams.
p. cm.
Rev. ed. of: Guerrilla gardening. c1983.
Bibliography: p.
Includes index.
ISBN 0-8007-7202-4
1. Fruit-culture. 2. Organic gardening. 3. Vegetable gardening. 4. Fruit—Varieties. 5. Vegetables—Varieties. 6. Fruit—Seeds. 7. Vegetables—Seeds. I. Adams, John Festus. Guerrilla gardening. II. Title.
SB357.24.A33 1988
635—dc19 88-d10060
CIP

Published by WYNWOOD™ Press
New York, New York
Printed in the United States of America

For Emily—again

For Susan Adams—a superb heirloom variety
and her grandfather
Spencer Adams—the finest guerrilla I've known

CONTENTS

LIST OF TABLES

GROW FRUITS & VEGETABLES THE WAY THEY USED TO TASTE

1. THE GUERRILLA GARDENERS

When the plant breeders at the University of California were requested to design a tomato for mechanical harvesting, they were instructed to think of the tomato not as a fruit but as a projectile. Consumers of the product they achieved can attest too well how successfully the breeders obeyed their orders. A recent educational-television feature explored further the relationship between the harvesting and shipping of tomatoes to the ideals pursued by the University plant breeders. The ideal is based totally on commercial convenience: tomatoes that are all ready to pick (not, of course, ripe) at the same time; tomatoes that can withstand maximum mechanical handling with minimum cosmetic damage; tomatoes that survive lengthy storage—three-month shelf life is now possible. There is even a breeding project to engineer a long cylindrical tomato that may be mechanically sliced into long uniform rows exactly conforming to the size and shape of a hamburger.

It is, of course, a cliché to say that "nothing tastes good anymore," and to add to this, "If you want food that tastes like it used to, you'll have to grow it yourself." Unfortunately, the solution of growing things for oneself is becoming only a limited option: of course, anything from one's own garden plot is going to be better than something that has survived the pipeline of commerce, even if the produce is grown from identical seed stock. But there, as Shakespeare said, is the rub.

Tomatoes—nearly all vegetables—have almost unintentionally been bred over generations of cultivation to fit particular requirements. Our ancestors, ignorant of theoretical genetics, kept seed from whatever produce best suited their requirements and needs or, more likely, most closely accommodated to a wide variety of requirements, some important and some less important. A good flavor in tomatoes, for example, augmented hopefully by the desirable feature of being early; augmented by the plant's being resistant to wilt; augmented by heavy production of large fruit. And so forth. The more desirable qualities that were being selected for, the more likely that they could not all be met by a single strain. So multiple strains were developed, and a specific gardener with space and time might grow several different varieties of the same produce to achieve different purposes. When a gardener began growing multiple varieties of the same crop, probably he began to obtain seeds from a commercial source to augment the seeds he saved from his own personal favorites. Since, in the beginning of the era of commercial seed houses, most of the varieties stocked were the same varieties he had been saving, perhaps he gave up perpetuating and saving his favorite seeds entirely.

When commercial suppliers became the usual source of gardeners' seeds and starts, new considerations began to determine what qualities in a fruit or vegetable would be preferred and its exemplars propagated. The number of varieties offered for sale certainly multiplied, but the multiplication was curiously qualified by a new kind or system of conformity. For one thing, the "morality of the marketplace" began to take over in the selling of the seeds as well as in the selling of the end product, mature fruit and produce. To a major extent appearance began to replace the hoary and venerable standard of flavor. Perhaps this replacement corresponded roughly to the arrival of catalogues with big, splashy color plates. While the big pictures seduced visually, debased language cooperated with empty promises of inner excellence: "delicious," "luscious," "succulent," "mouth-watering," the pitchman lied. The word list grows correspondingly longer as the quality they synthesize degenerated to fainter and fainter.

While many seed catalogues continue to list varieties that are sold for genuine quality and not following the lead of commercial marketing, the tendency has been for listings of such kinds to become smaller and smaller. Some seed houses that pride themselves on their listing of older varieties remain, and some small catalogues are devoted to such items exclusively. For the most part, however, the large and colorful catalogue most gardeners receive each January will be found to be listing fewer and fewer of such noncommercial varieties each season.

In recent years an additional factor has accelerated this disappearance of traditional varieties from seed catalogues. There has been a tendency for smaller seed companies to be swallowed up by larger companies, with correspondent limiting of seed varieties offered the retail trade and of the stock grown to be supplied to commercial growers. It is commercially more efficient and practical to grow one variety for all seed markets. Another trend has been for the larger seed companies to be taken over by large corporate conglomerates. The Graham Center in 1979, for example, lists fifty-seven acquisitions of seed companies, including Burpees by ITT, Northrup King by Sandoz, Dessert Seed Company by Atlantic Richfield, Keystone Seed Company by Union Carbide, Ferry Morse by Purex, etc., etc., etc. Such takeovers fit current patterns of diversification by major companies, and while we may be sure they must be good for business, as gardeners we are fairly sure they are not likely to be good for us.

The absorption of small companies by large, and the takeovers of companies large and small by corporate interests, tend to result in seed offerings by such companies being reevaluated from a perspective that is strictly commercial. Seeds that don't sell well, like less profitable lines of products in any commercial venture, will not be offered long. And again, what is offered to the public becomes more and more restricted to what is sold to commercial agriculture. With such restrictions, the traditional wisdom of the marketplace prevails: if we don't have what the public wants, let's make them want what we have.

A frequently cited example of how commercial images dominate and control what gardeners buy as well as what is most easily available to them is the dominant place occupied both in produce markets and in nursery suppliers by the red delicious apple. Without engaging in the controversy over how good they are or aren't, at least they lack the versatility to be, as to a degree they have become, the universal centerpiece of home orchards. In some (rather restricted) areas and climates they grow very well; in some areas where they are nevertheless extensively grown they do poorly. As good as they arguably are or are not, there are other varieties which almost unarguably are better for most of the purposes for which the red delicious is grown. However, few people will dispute that the red delicious is an absolutely beautiful fruit. That, of course, is what sells it. With its same flavor, textures, keeping qualities, versatility, etc., if in appearance it were unsymmetrical, pale yellow with pinkish blotches with occasional small horny worts, the apple would be hard to give away. Yet the apple whose appearance I have described might be one of the several cultivars of the pippin, an apple favored in England since at least the sixteenth century, and an apple called by many the best-flavored in existence. Obviously such an apple would attract no notice in the supermarket and would also be passed over by writers of promotional material. And since it would have no recognition value in a nursery catalogue, it would have (and does have) little recognition value for the home gardener. Therefore, the pippin apples have been virtual strangers to patrons of most commercial-nursery catalogues for fifty years or more. Much the same circumstances condemn the greengage plum to similar obscurity, although adding to its lack of visual appeal, the ripe fruit is so delicate that shipping it is out of the question. Yet the greengage and its cultivars are considered by experts of taste the standard by which all other plums should be judged in tasting sessions.

For several years I have looked for a variety of apple called "Arkansas black." It is a "memory apple"—"Haven't seen it since I was a boy," as someone is always saying. In my memory it is the most beautiful apple in existence. It must be the model of the poisoned apple of Snow White, or of the Apple of Discord in Greek mythology: perfectly round, huge, and with an unflawed skin as red-black as a dead-ripe bing cherry. The reason it is not among the royalty of commercial apples is that it is not only as large and round as a cannonball, it is nearly as hard. It becomes edible only in late winter. I wanted one because it is so gorgeous, and to have one would be a memory trip.

As it turns out, I must not really have looked that hard or I would have found it. It is not handled by any local nurseries I know, and the usual handful of commercial seed catalogues I am familiar with does not begin to represent the variety of seed and plant sources, small and not so small, that can be found with a little looking. And most of all I had not realized how far I was from being alone in my interest in antique varieties of fruits and vegetables. That it is not really a fraternity of people of such interests is only because many are not aware of the existence of others. If they lack a suitable uniform *pronomen* or distinctive name, there is a vital center of people who designate their common interest by calling themselves quite insufficiently "seed savers."

Essentially this is a largely unaffiliated group of guerrilla gardeners who live and function somewhere outside the constraining and conforming pale structured by the conventional seed catalogue. While they do not disdain commercial sources (and some of these commercial sources, in spite of the national trends, are very good indeed, a few almost qualifying to be considered as guerrillas themselves), this group of gardeners is dedicated to growing as well as preserving old varieties and rare varieties of every sort of crop the earth supports. Essentially people with like interests, they exchange mostly on a barter basis or an exchange of postage various seeds which they save or various fruits and nuts which they propagate.

Only in a misleading sense of the word can these guerrillas of horticulture be called a group. There are organizations to which they belong, and publications and registers in which their names and interests are printed, but at least as yet they are not particularly groupy. They are generally free of hang-ups or orthodoxies or in-groupish affiliations, although certain clubbish groups seem in some cases to have coalesced.

There is no standard terminology for the object of their exchange or standard of definition of the objects of their propagation and exchange. The expression "old-timey seeds" is sometimes encountered, as is sometimes "old-fashioned varieties." What seems to be the most prevalent term, however, and the one most likely to stick, is "heirloom varieties," although even this term may not be sufficient to indicate the range and kinds of interest. Besides old-fashioned or "memory" varieties, they collect and exchange unusual varieties, varieties not usually encountered in home gardens, varieties generally unknown but possessed of unusual virtues, varieties which extend the usual climactic range of a given species, etc.

The meaning of the term "heirloom variety" is necessarily indistinct. It almost can mean what an individual wants it to mean. If you're eighty years old and remember a bean your grandmother used to dry in Arkansas, that's an heirloom variety. If you're thirty and remember a green bean your grandmother grew in Oregon which you have not seen since but would like to find, that's an heirloom, too. If you're looking for a variety of sweet corn you haven't seen in a catalogue since 1950, that's an heirloom. By and large, however, for purposes of general definition there are several more or less arbitrary benchmarks which affiliates of the guerrilla band take into account. By one reckoning, any plant variety in existence before World War I may be called an heirloom. Some people offer the turn of the twentieth century as the line of demarkation; some the middle of the nineteenth century. For purposes of dating, old seed catalogues, gardening books, technical books, agriculture-college publications, etc., are valuable sources of reference. Such publications are also valuable in helping to identify unknown or uncertain varieties.

There is an additional category of heirloom varieties with more precise, and certainly more romantic, dating. These are the pre-Columbian varieties: varieties that were grown in the New World before the first voyage of Columbus. On first hearing a reference to "pre-Columbian corn," for example, the initial reaction of many is to interpret the term to mean that the seeds were found in ancient tombs. Sensational stories to the contrary, most members of the scientific community are skeptical of all claims concerning growing seeds recovered from ancient tombs, in the Americas or elsewhere. The existence of pre-Columbian seeds is explained much more simply. The numerous varieties of seeds grown before the voyages of Columbus continue to exist because they have been grown continuously by the descendants of the pre-Columbian people in the Americas who developed them in the first place.

The pre-Columbian list is a long and varied one, for it includes examples of most of the plants the New World gave to the Old. Included are varieties of corn, squash, beans, peppers and potatoes, as well as such lesser-known native foods as the Jerusalem artichoke and the ground nut. Almost conspicuously lacking from the list of pre-Columbian plants is the tomato. It is also a New World plant, but it seems to have been little grown by the Americans, although such cousins as the ground cherry and the tomatilla were cultivated to a certain extent. Only one tomato species, the ancestor of the various small "patio tomatoes" of modern cultivation, seems to have been grown in a limited way. The tomato came into its own in varieties developed after they were introduced to Europe as novelties, probably primarily as ornamentals.

Why individuals want to locate and grow heirloom vegetables is fairly clear and perfectly understandable: to rediscover and propagate those lost memory crops, fruits and vegetables that fulfill in taste the real or imagined potential of that particular species and avoid the disagreeable qualities, real or imagined, that were bred into them or out of them in the pursuit of commercial considerations. Ultimately most heirloom gardeners would say in effect that they want to make the garden taste good again.

The reason for growing pre-Columbian crops is similar, but with a difference. The oldest among us, of course, is hardly likely to have memory trips for pre-Columbian things eaten in childhood. Nevertheless, flavor can certainly be a consideration. The pinquito bean, for example, is the nonpareil bean for chili. It can be found in stores that cater to ethnic groups which know it from Mexico, but the surest way to have this bean is to grow it oneself. The tepary bean, grown today by a small number of Indians in restricted parts of the American Southwest and northern Mexico, is among the most drought-resistant crops known, and it is also nutritionally superior to common varieties of beans; these are both sufficient justifications for giving it a try. Pueblo blue corn—pre-Columbian also—is the *crème de la crème* of tortilla corn. Homemade tortillas from its meal are considered by some as much superior to commercially supplied tortillas as garden tomatoes to machine-harvested commercial tomatoes. And it shouldn't matter to the connoisseur that the tortillas themselves are colored a mild but unmistakable blue.

Numerous practical reasons can be advanced for growing, or at least trying once, specimens of numerous pre-Columbian varieties. For a great many gardeners, of course, reason enough to try at least some of them is that they are there. If I had known five years ago that such seeds could be obtained other than from university seed banks and horticultural collections, nothing could have kept me from growing some.

There are soberer and less whimsical reasons and considerations to propagate and perpetuate all manner of heirloom varieties besides personal pleasure. There is always the possibility that all or some of these varieties will simply cease to exist if they are not propagated by individuals and distributed through the guerrilla underground. There has been a long-range trend among large seed companies to drop less popular and hence less profitable varieties of plants from their production and lists. As seed companies and nurseries are increasingly taken over by large corporations, this tendency, already far advanced, is expected to increase.

Another trend that threatens traditional seeds is downright Orwellian. Although individual companies do develop and claim ownership of new varieties, basically, except for some ornamental varieties such as roses, plants and seed have always been, in whatever variety they grew, public property. In recent years there has been a movement to pass laws all over the world, but especially in Europe and North America, that would effectively take ownership of plant species out of the public domain and make such ownership private and commercial. Because the subtle differences between similar varieties of plants makes legal distinctions extremely difficult, there is a movement of members of the European Common Market to restrict varieties of plants that can be grown to those listed in the "Common Catalog." Dr. Erna Bennett of the United Nations Food and Agriculture Organization has predicted that within ten years of the time this system goes into effect, three quarters of the vegetable varieties grown in Europe will be extinct. The effect of the "Common Catalog" listing can be not only to establish a monopoly on seed varieties commercially available, but to make the growing of nonlisted varieties unlawful. It is certain that this plant-patenting law makes seed companies more commercially attractive. Within a week of the passage of the Common Market law in England, one corporation bought eighty-four seed companies, and it currently owns more than a hundred. Those supporting such laws claim that by making ownership of a plant variety profitable, the laws encourage research and development to produce patentable plants. The horticultural community fears just the opposite kind of result. With decreasing genetic diversity supplied by multiple varieties, species can become vulnerable to the point of being bred out or even becoming extinct.

To add to the Orwellian scene, a similar law was signed in the United States by President Nixon in 1970. As a harbinger of what American gardeners should expect from commercial seed houses in the future, the 1983 *Burpee Gardens Catalogue* (ITT) designated some eleven varieties of vegetables "Unauthorized propagation prohibited—U. S. protected variety." Technically, individual gardeners may not save seeds from these varieties. In at least two instances, the protected variety replaced in the catalogue a variety no longer listed; as, "Oak Leaf—Discontinued. We recommend Royal Oak Leaf," a U. S. protected variety (p. 106). We may guess this substitution of protected varieties for non-protected is a trend, foreshadowing future applications of the provisions of the plant protection act. It will not occur overnight, and probably it will be another decade before it will begin seriously to restrain the access gardeners have to seed, but the pattern seems clearly established; the concern of the seed savers that old and open pollinated varieties may cease to be commercially available is not hypothetical.

It is hardly conceivable that Americans would ever stand still for the prohibition and proscription of any plant varieties, but at least in theory America will soon be the only place where three fourths of the plants cultivated in Europe may legally be grown. If something similar happens in America it will first have to account to the guerrilla gardeners.

The restriction of genetic diversity is potentially disastrous. Keeping old varieties alive sustains potentially valuable genetic differences. The larger the gene pool for a particular species, the more that species has the potential to produce variations of itself. With a wide variety of differing genes to work with, the possibility increases for breeding genetic configurations that are, for example, tolerant of certain soil conditions, resistant to heat, cold, drought, etc.; resistant to disease and to fungal infections; resistant to certain insects. All of these qualities plus many more have been bred into individual species.

In some ways cloning—growing from root cuttings, tuber cuttings, corm division or any asexual system of reproduction—might seem the ideal way of reproduction. You have the plant you want, and can reproduce it *in perpetuity* without change. However, asexual reproduction rests on a precariously narrow genetic base. Bananas, for example, are reproduced by cloning so exclusively that almost all have lost the ability to produce seeds. Because of their limited genetic pool, there is serious worry about the possibility of maintaining a banana industry indefinitely. Diseases and blights that have developed in recent years are a serious threat, and one which seems to be without a genetic answer.

Very close to home, and recently, the American corn industry had a close brush with disaster. Virtually the entirety of the U. S. field-corn crop consists of hybrid varieties of hard dent corn. The seed corn is produced by cross-pollinizing an inbred strain with another selected variety. The result is a corn having numerous desired characteristics. If planted for seed, however, this corn crop would not produce corn like itself the following year, but a different and quite inferior corn. Therefore the seed for next year's corn crop must be produced by repeating the original cross each year. Such hybridizations are among the crown jewels of modern genetic achievements.

But there are potential hazards. In 1970, a blight developed in America's hybrid-field-corn crop, moving from the south ern areas north through the heart of the corn-growing Midwest. In the worst year of the blight nearly 50 percent of the nation's corn crop was lost. Had there been no genetic reserves besides the varieties being used to breed the hybrid seed, theoretically at least corn could have ceased to exist, at least as a practical farm crop. Fortunately there is a large pool of corn germ plasm deliberately preserved, and this unforeseen disaster confirmed the logic and necessity of such preservation. From these reserves and resources a new variety of corn was developed that was resistant to the blight.

An analogy to what might have happened to corn might be the mung bean or chick-peas. Both are very restricted species with very small gene pools. With almost no genetic variety to work with, a geneticist would have almost no chance of developing new genetic combinations that could survive. Similarly, the soybean has been cultivated since ancient times, especially in China. In the twentieth century, however, the selection of a few strains of commercially grown beans has caused the loss of incalculable numbers of varieties once locally grown. Geneticists are alarmed at the resulting shrinkage of genetic diversity. What happened to corn in 1970 could quite conceivably be repeated with soybeans. Consequently a priority of soy research is the seeking out of old or wild varieties that may still exist in China and elsewhere. A deliberate aim of some groups of the seed savers is to preserve just such genetically diverse varieties.

Some of the most dedicated of the guerrilla band who save seeds and husband old varieties have developed remarkable private and community collections. Some have been accumulating and maintaining such collections for many years. Perhaps the most famous and most remarkable of these is the so-called Wanigan Associates Collection. Assembled and maintained personally by Mr. John Withee, this collection includes over 1,200 varieties of heirloom beans. Because of ill health, Mr. Withee is no longer able to personally maintain his collection—bean seeds must be renewed every four years if the strain is to be kept alive—and it has been taken over for maintenance and propagation by Mr. Kent Whealy, president of the nonprofit corporation calling itself the Seed Savers' Exchange.

The Seed Savers' Exchange is dedicated basically to the discovery and perpetuation of heirloom plant varieties, and is helping sustain small and large private collections that might be in danger of disappearing, as well as providing limited accessibility to these collections. The exchange is also the major coordinating force of the guerrilla gardeners. It and its publications are the major avenue through which large numbers of enthusiasts make contact with each other, discover sources for varieties they need or want and offer to other growers varieties they propagate.

Members of this group are listed in the organization's yearbook and directory by state, with the plants and seeds they have available for exchange as well as plants they are interested in obtaining. A word of sober caution: neither this directory nor any of the seed lists of the various participants in seed exchanging should be considered seed catalogues. They and the Seed Savers' Exchange are not in any way commercial enterprises. To treat them or make use of them as if they were commercial services could severely damage the practical application of the seed exchange concept. Those seriously interested in heirloom seeds and rare or unusual varieties may easily become members by paying a small fee and observing the simple and sensible rules for offering seeds and making use of the exchange.

There are a number of commercial or semicommercial sources which have more or less limited listings of heirloom seeds and plants or otherwise rare and unusual plants. While these may be considered business firms, they are often very small-scale operations, and usually very personalized in their services. Some charge for their catalogues, but even if they do not, a courteous person when writing for their catalogues (or more usually, merely seed listing) should enclose a couple of first-class postage stamps.

Among the most respected of these small seed houses is Johnny's Selected Seeds. Although commercial, it decidedly favors the interests of those who propagate and save their own seeds. A person whose interest is merely to find some older variety, and who doesn't particularly know even what he's looking for, could hardly find a better place to begin than this and, unless he really gets the bug, probably would never need to go further. More specialized, *The Vermont Bean Book* provides specimens of a dizzying variety of beans, with decent descriptions of what they look like and what they are good for. The smaller *Good Seeds* is a nifty personally designed catalogue that offers a fine selection of old and unusual varieties. It has a fine offering of pre-Columbian species. For nursery stock *Henry Leuthardt Nurseries* has a small listing of a number of old varieties of fruit. I have found their service impressively personal, and the nursery stock I have ordered from them to be top quality and perfectly packed for shipment. The best of all for range of selection of old fruit varieties is *Southmeadow Fruit Gardens.* Not only is their range and variety of fruits awesome, but the descriptions of varieties as well as of their background and family trees is authoritative and thorough. Also the text is literate and a pleasure to read.

These are but a tiny number of commercial or more or less commercial houses specializing in offering heirloom varieties. Addresses for these and prices, where applicable, will be found in Appendix I. Names and addresses of noncommercial plant sources will be found in Appendix II. Sources in Appendices II and III will give access to becoming acquainted with the purely private collectors and exchangers of heirloom and endangered species. The first step to cutting loose from the roster of the regulars and becoming a guerrilla gardener is to stop thinking of conventional crops and of conventional ways of coming by seeds and plant stocks. Stop thinking of supermarket tomatoes and apples and beans. Fruits and vegetables grown for reasons other than such commercial advantages as appearance and ability to withstand shipping are still available, for a while at least. With enough gardeners interested in preservation, they will be around for a long time.

There is one lesson about endangered species that should be remembered if one decides to become involved with the brotherhood of seed exchangers. The oldest known living things are the bristlecone pines. In 1964 a specimen in the Sierras was determined to be 4,900 years old. Although its location was kept as secret as possible, a short time after it was studied someone cut it down with a chain saw. The guerrilla gardeners, seed savers and seed exchangers are also a rare and precarious breed. Use them accordingly.

2. THE GUERRILLA IN THE VEGETABLE GARDEN

The possibilities for experimenting with heirloom vegetables are endless. There is certainly no necessity for becoming totally hooked on growing exclusively antique varieties, although the temptation to do so overcomes many guerrilla gardeners. Very probably, on a small scale at least, most people who become interested in even a single heirloom variety or uncommon species will become seed savers. And from being one who saves seeds for his own garden, it is another easy step to becoming a seed exchanger.

By its very nature, the hobby of searching for new/old seeds and exchanging them with others makes a definitive listing of seeds and varieties impossible. Reading the publications and listings of the guerrilla gardeners shows emerging not a consolidated list of possibilities, but a baroque movement of ever-increasing possibilities. Even if the possibilities were capable of being exhausted, the arrangements, formations and exchanges possible indicate infinite recombinations and reunitings. It's as if the expanding range of possibilities were itself a kind of gene pool, and the larger and further it expands the more combinant possibilities exist. And the more possibilities of retaining a vital and healthy strain of gardeners.

It would be impudent to offer even a list of "suggested" offbeat or heirloom varieties of vegetables. Nevertheless, it may be instructive, and it might to a certain extent point to the range of possibilities, to describe with some minimum family background a certain number of plants which some of the guerrilla gardeners are interested in. Seeds for all of these varieties are obtainable from commercial or semicommercial sources listed in Appendix I. I have also seen all of these varieties, and, of course, vast numbers more, listed or offered by seed savers.

Although for various reasons it may not be the most practical crop for the home gardener to experiment with, corn strikes me as the most glamorous of the heirloom vegetables; other people, for their own differing reasons, would likely choose otherwise. Tomatoes are fine, wonderful, and would probably be the last open-pollinized vegetable I would willingly give up. And the tomato certainly is not without its own romantic past. But ancient corn, pre-Columbian corn, carries about it an aura of wildness and certainly of mystery. With its strongly colored and oddly shaped kernels sometimes irregularly placed on the cob and its total unlikeness to hybrid dent corn, it evokes an immediate sense of the past that is almost prescient. The cultures that originate most of this corn are long dead, but the grain still evokes a sense of their living reality.

In that it exists at all, corn is a mystery; it is, in fact, an evolutionary anomaly. Large kernels growing around the circumference of a handy-sized cylinder and protected by thick tough leaves makes about the tidiest package nature can offer humans. Wheat and other small grains require for a primitive farmer laborious collection to provide a meager number of small grains, then difficult separation from the other plant materials (quite laborious and difficult in the case of barley, oats and rice) and careful storage, plus a variety of repetitious steps, with a relatively small return for the effort. By comparison, corn seems almost to have been engineered specifically to be handy for humans.

It is certainly not perfectly engineered for survival. According to evolutionary theory, every change that occurs in a plant or animal is retained only if it improves that organism's chances of survival. All of the features of corn which make it desirable for human use inhibit its survival and propagation in nature. In fact, corn, even in the most primitive forms known, is presently incapable of propagating itself. If left to reproduce naturally, corn would become extinct in only a half-dozen or so years.

Because the form in which corn grows contributes no survival value—in fact, they all inhibit survival—it would seem that it is a form developed under cultivation, since cultivation provides guaranteed propagation. Two new problems or curiosities appear, although neither is unanswerable. Since natural selection would not have favored it, the handy form in which corn exists implies that the early plant breeders were selecting toward achieving that form. Since no analogous form exists in nature, it is hard to escape the conclusion that these early men were thinking the unthinkable, exercising a prodigious leap of the imagination.

The second anomaly about the development of corn is that its primitive wild ancestor has never been identified. Numerous possibilities, many being plants having hardly the slightest resemblance to corn, have been examined. None studied, however, has the necessary genetic structure to have been the ancestor. Teosinte, a sometimes cornlike plant of Mexico, has been a longtime candidate for this ancestry. Some promising leads have proven to be backcrosses from cultivated corn, rather than the other way around. A recently discovered unique specimen of teosinte again raises possibilities of tracing the ancestry of corn to that quarter.

Aside from satisfying innate curiosity in answering the question, there could be practical application as well if the actual ancestor could be identified. It could provide a completely new genetic pool, enlarging further the possibilities of corn breeding.

In spite of the fact that the majority of garden and commercial sweet corn as well as virtually all of American field corn are hybrids, enough varieties of heirloom corn survive to provide a lifetime of experimentation. Quite a number of varieties, including pre-Columbian, are now available through some commercial and semicommercial sources, and of course many more are propagated by the guerrilla gardeners. In addition to those which are available through a little looking, an almost unbelievable number of varieties are preserved in collections of various universities and foundations. For example, the Rockefeller Foundation and the National Academy of Sciences maintain at the University of Mexico City a collection of some twelve thousand specimens. This may be the largest collection that exists anywhere, but there are other collections maintained by North American universities as well.

However it originated, it is likely that corn was first cultivated somewhere in South America. From its original home it migrated, apparently rather slowly, both north and south. It seems to have arrived in the upper half of North America relatively late. Perhaps varieties were slow to adapt to the colder growing conditions and the shorter season.

Among the heirloom corns of northern North America are varieties associated both with the Indians of the North and of the American Southwest. The exact heritage of some heirloom varieties of northern corn is unknown, but they are supposed logically to trace back to corn originally obtained by the first settlers from the Indians. Certainly they introduced those settlers to corn and to the methods of cultivation. The variety called old red, or old red field corn, is one such variety. It was commonly grown in Colonial America, very likely obtained from the Indians in the same form in which it is presently known. It may have been the corn of the first Thanksgiving. Another surviving Colonial variety, familiar in the region farther south, is called hickory king.

There are fewer corn varieties associated with Northwestern Indians than with either Eastern or Southwestern. A missionary with De Soto noted in his diary that an Indian had told him corn would not be found west of the Mississippi River because of the herds of buffalo. That makes sense. What one run-a-gate cow can do to a planting of corn is sufficient to make a gardener weep. We can suppose that the agriculture of the Midwest would be quite different today if herds of buffalo still migrated the plains between Chicago and Oklahoma City.

There is a corn variety called Mandan bride associated with

Indians of the Dakotas. I have grown it once. It requires a relatively short growing season, and at least in my crop the stalks were of a medium to short height. The ears were generally small, six inches or less, with the kernels both larger and more rounded than common dent. The kernels are usually variecolored, black, white, red and shades in between. Eaten as green corn it had a rich and pleasant "corny" flavor, lacking most of the sweetness associated with garden sweet-corn varieties. Ground into coarse meal, it made a richly flavored corn bread.

Far more varieties of corn are associated with the American Southwest. Some of these varieties are particularly valued for the *masa* flour of Southwestern and Mexican cooking. Some varieties are uniquely tolerant of drought. I tried one such drought-resistant variety, a pre-Columbian corn called Hopi blue, which is said to be among the most drought-resistant varieties known. I planted it in an especially dry spot that had never before been cultivated, determined to see what it would do with no water at all. The area is also covered with a layer of volcanic ash from Mount St. Helens about ten inches deep. During a hot spell in late July it began to show distinct distress, and I decided that the handicap of the ash was perhaps too much to ask the corn to overcome. I applied a side dressing of fertilizer and watered it once. Perhaps as much a result of the fertilizer as of the watering—the ash has zero nutrients, and the roots must reach the real soil below to find nourishment—the corn recovered and made good growth. Because the corn requires 105 days to mature, which it did not get, nothing matured, although it did grow fine long ears.

Another pre-Columbian corn from the American Southwest is called Pueblo blue. It has the reputation of making superior tortillas, although it is said they will be blue in color, like the corn from which they are made. In my planting the stalks were quite short, and they were not very productive. Eaten green the corn had a rich corn flavor, and even immature it seemed quite tough. A neighbor's cow visited the patch as it was ripening, and it will be another season before I know if it really does make blue tortillas. Of the few ears that survived, the rows were straight and uniform, and the kernels also uniform and evenly spaced. The color was a rich blue.

One of the pre-Columbian corns that is an actual possibility (as distinct from a curiosity) as a green corn is the black Mexican. It has a flavor that many people find exceptionally attractive, although it would hardly be described as sweet. A three-year-old girl who was handed a raw ear in the garden for sampling quietly ate the cob clean.

A corn that long has piqued my curiosity is the miniature corn sold pickled in bottles and found as an ingredient in Chinese cooking, etc. Nowhere had I ever heard this corn or its cultivation discussed, nor seen its seed offered for sale. Finally I located seed under the name "baby corn," offered by Sunrise Enterprises. I was mildly surprised to note the date to maturity, in excess of one hundred days, which seems rather a lot for such minuscule little ears. It was also noted that each stalk produced four or five ears. Germination for me was very poor, no more than five plants in a ten-foot row. But those five grew in a most astonishing fashion, and by the end of the summer I realized why they had such a late date of maturity. By the time tassels appeared they were gigantic, over eight feet tall, and with their tassels, which were pendulous rather than erect, they must have topped ten feet. The stalks were not only tall, they were thick and rugged, and the leaves were on a scale to match. One plant set two ears, but ran out of growing season before they matured. Inside great wads of corn shucks were indeed tiny little ears, even though the kernels were not pollinized or fully formed. I have since discovered that the tassels should be removed before they have mature pollen, and the ears harvested as soon as they develop silk. This is a variety that the inveterately curious will probably be compelled to try.

There are a number of traditional or heirloom varieties of sweet corn available commercially, and many more are grown by the seed exchangers. I have tried to grow a few of the old classics such as Stowell's evergreen, country gentleman and other shoe-peg varieties with little success. They thrive in a climate of higher humidity than my area and also enjoy warmer nights and especially good long growing seasons. Where they may be grown these varieties are said to be truly superior.

There are a great many open pollinated heirloom varieties available for experimentation, as reading the Seed Savers' Exchange, lists provided by guerrilla gardeners, and even good commercial catalogues will attest. However, as Rob Johnson notes in his catalogue *Johnny's Select Seeds*, it is unmistakably true that none of these old varieties are as sweet as modern hybrids. On the other hand, sweetness may have become overrated as the single criterion of superior flavor.

Another New World crop that offers nearly limitless possibilities for experimental culture is the beans. The possibility of becoming hooked as a bean collector might seem remote to one for whom the idea of growing noncommercial vegetables is still novel, but the existence of such accumulations as John Withee's 1,200-plus varieties confirms that it can happen. Looking through the interests of gardeners registered in the Seed Savers' Exchange reveals many more "serious" bean collectors. For some people alternative plant varieties and alternative seed sources have become almost a way of life.

For the average individual, however, it is satisfaction enough to learn that one is not restricted in choices to commercial varieties, and that even with such a thing as dried beans, where, one suspects, the humble nature of the crop itself makes variety unimportant, not only does variety exist, but there are absolute surprises as well. Because one expects so little from dry beans, most gardeners are content to buy what beans they have need for, with little question of trying to grow their own. Although occasionally a catalogue offers as many as half a dozen varieties, most offerings that include dry beans give pretty much the identical varieties to be found in a grocery store. A few commercial catalogues, however, offer considerable variety, but to truly comprehend the range of possibilities one must look at the varieties offered by the seed exchangers.

All New World beans are distinct from the Old World bean, the fava; they are native only to the Americas, and it seems that at the time of Columbus' voyages they were pretty well distributed along the length of the two continents. There are four distinct species of American bean. The largest family is *Phaseolus vulgaris,* the common bean, which includes both the dry and the green varieties. The distinct family of the lima bean is the *P. lunatus.* The scarlet runner bean, although it appears to be no more than a particularly vigorous climbing *P. vulgaris,* is a member of a distinct family, the *P. coccineus.* The fourth species is the *P. acutifolius,* commonly called the tepary bean.

Some varieties of *P. vulgaris* are grown for both green and dry use. Some of those grown exclusively as dry beans have even in their most immature stage too much fibrous material to be satisfactory as a green bean. All varieties grown to be eaten green may be used as dry beans, although some may be too mealy to be palatable.

Among the heirloom varieties the Vermont cranberry bean is well spoken of and is of excellent quality both green and dry. It is also early, coming off as a green bean in sixty days. The immature pod is somewhat mottled red-green, but it cooks to a basically pale green. It has a fibrous string which must be removed from both the top and the bottom of each pod before cooking. It has a fine and distinctive flavor when picked at the proper stage, but as the beans begin to mature the pods become too fibrous to be eaten. At this stage the immature beans, shelled, are excellent. Some declare this bean to be the finest of the so-called shell beans. When dried, the beans have a mild flavor and seem to be especially easy to digest. The Vermont cranberry bean has been grown in New England for generations, and it may be pre-Columbian.

The red Mexican is a pre-Columbian bush bean which matures in eighty-five days. Although I haven't tried them as a green bean, the hulls are not especially fleshy, and I do not think they would be particularly suitable. The plants are highly productive and are especially easy to shell. I have cooked them only with chili for which they were fine, but there was nothing to distinguish them especially from other red beans I customarily use.

The pinquito, or Santa Maria pinquito, from the principle place of cultivation, is a bean of distinction in chili. Some people honor it as the preeminent chili bean, the gourmet ingredient of that humble dish. They require a moderately long growing season, about ninety days. I had trouble getting them mature enough to dry and harvest, and many of the beans molded rather than dried. I had been led to believe that the pinquito grew as a bush bean, but it is somewhere between a bush and a climber, perhaps more accurately designated a "sprawler." It refused to be a bush but declined to climb when offered support. The beans themselves are very small, scarcely half the size of a common red bean, and they are wrinkled like some peas rather than smooth like most beans. Pinquitos are exceptionally low in starch, and they are noted for not breaking up while cooking. My plants set pods prolifically, but because of the smallness of the beans and perhaps because there were relatively fewer beans per pod, they were not especially productive. They are, however, excellent in chili, and they make the bean a distinguished partner of the dish rather than merely an ingredient. Some recipes call for them specifically. Although I have seen them for sale in stores catering to ethnic trade, for the most part if one wants them one must grow them for oneself. Probably most people who grow them once as a curiosity will grow them again for their own sake.

Although not a pre-Columbian heirloom, the French flageollet, or flageollet vert, is a rarity in American gardens and American seed catalogues, and it is a bean that deserves to be more widely cultivated. The name is seldom mentioned without adding that it is *the* French gourmet bean, with sometimes specific mention of its use with lamb. The immature bean hull is pale yellow, and the mature dry bean is white and slender, the size and shape of most common green beans. It grows as a compact bush, and while it is sometimes described as requiring one hundred days to maturity, it is in good form as a green bean much earlier than this. I found them superb both as green beans and as shell beans, and everyone who tasted them gave them the highest praise. They have a delicacy of flavor, as well as a mild distinctiveness of flavor, that is unlike other beans. They also have a curious feature described by some as "making their own sauce." Even green, there is a thickening of the cooking broth as if they were indeed being cooked in a specially prepared sauce, and this feature becomes more pronounced when they are cooked as a dry bean. While the flageollet is associated with French cooking, it should not be relegated to a category of limited use and rejected on that basis. I would like to try it in French recipes, which I have not, but it is excellent in its own right in the ways green beans are normally cooked. It is not, in other words, an exclusive bean or one with limited or restricted uses. It is a good general-purpose green bean.

The Oregon giant bean is an heirloom variety that has been associated with western Oregon since the earliest pioneering days. It is a large bean in every way. A climber, it wants lots of height because it grows tall. The long (eight inches plus) pods are mottled with pale blue, and they have a tough string that must be removed. As they begin to approach the shelly-bean stage it is best to remove the string from both the top and the bottom of the pod. The pods are so large that a dozen make a family serving. In spite of their great size, they cook quickly and are very tender. The flavor is mild but good, with distinctly more character than many of the popular commercial green beans. They are not particularly distinguished when cooked as a dry bean.

The number of varieties of common beans which are available is much too large to précis. Such catalogues as *G Seeds, Nichols, Johnny's* and *The Vermont Bean Book* each offer respectable selections of heirloom and pre-Columbian varieties and also give excellent descriptions of the distinguishing qualities and particular uses of what they sell. In addition, the seed savers offer and describe numerous more varieties. Even a partial listing of the Wanigan Associates (Withee) collection in the *Seed Savers' Exchange* begins almost to numb the mind.

The tepary bean (*P. acutifolius*) comprises a small but potentially important species of bean. How many varieties there are within this species does not seem to be clear. It is a pre-Columbian bean apparently limited in its native range to the deserts of New Mexico and Mexico. It has three qualities that are particularly interesting. It is very drought resistant, is very heat resistant and has nutritional values that are superior to *P. vulgaris.* Grown with only natural rainfall in the Southwestern deserts, it is described as literally growing itself to death when rainfall did occur. I grew it in 1982 with mixed results. It germinated very poorly, which was perhaps because of the shocking difference in spring warmth in my area in contrast to the bean's desert homeland. The ones that did germinate were perhaps overwatered in terms of their expectations. The plants grew as sort of crouching bushes, neither truly bushes nor truly vines. They spread into a somewhat dish-shaped configuration about four feet across. The foliage was a grayish green, and while there certainly was a beanlike look to their shape, they still looked distinctly different from common beans. The pods were small and shaped more like a soybean pod than a common bean, and there were only two or three beans per pod. Production was light; there were not enough to experiment with cooking. I remain bullish on teparies, and will try them again; I think there may be a future for them.

There are several varieties of pre-Columbian lima beans (*P. lunatus*) that are relatively easy to obtain, and many more rarer varieties are available through the resources of the seed savers. Some varieties require exceptionally long growing seasons. In 1982 I grew a variety called the Hopi mottled, a pre-Columbian lima of the Hopi Indians. In the catalogue from which I ordered, they were described as being a pole bean bearing four-inch pods and requiring one hundred days for maturity. I planted them with tripod trellises, and when they germinated I waited for them to begin to climb. They grew and spread (but not vigorously until the truly hot days of summer), in a growing habit much resembling the tepary bean, but they resolutely refused to climb. Despite my help, even my insistence, they refused the support of the poles and continued to sprawl across the ground. They set pods perhaps two inches in length, but they were rather scant. Except for their refusal to climb, most of their unimpressive performance is probably attributable to the local climate. I have never had favorable results with any lima beans, even those specifically adapted to short seasons.

Although it is certainly not an heirloom variety, I have also tried growing the edible soybean, so called because it may be eaten immature, like a green or lima bean. In fact, I hoped it might offset the difficulty I have had trying to grow lima beans and might serve as a replacement for them. The bushes grew well, in fact a great deal better than I had expected. They were not sufficiently productive, however, to interest me in trying to grow them again. Each pod contained two beans and, for the rather uninspiring results, seemed hardly worth the effort of shelling. Other people, valuing the quality of the soy's protein, might judge differently. Or in a different climate the production might be different.

Going below ground, the Jerusalem artichoke is a pre-Columbian vegetable which is beginning to become widely known only some five hundred years after the discovery of America. Part of the recent interest in this vegetable comes from the fact that the starch of its tubers is compatible with the dietary restrictions of diabetics, for whom it can essentially become a potato substitute. The same reasons make it of interest to a wide range of diet-conscious people. In recent years it has become relatively common seasonally in general-produce markets.

The tuber of the Jerusalem artichoke is rather knobby, resembling a number of large carrot tips joined or lumped together. The skin is white to yellowish white, and the flesh is much the same color as the skin. The flesh is very crisp, and when raw it adds a pleasant texture to salads. The flavor is somewhat off-sweet, and not especially distinguished. Besides being eaten raw, Jerusalem artichokes are cooked in most of the ways potatoes are cooked.

Although until recently little grown, the Jerusalem artichoke was an important food crop of North American Indians. Places where there is a heavy local concentration of the plants often identify sites of ancient Indian encampments. Since they tended to grow around the midden heaps of old camps, Indians returned to the same encampments from year to year to harvest them. They are thought to have provided a link from the food-gathering society to the food-cultivating society.

There is some confusion about the name of this plant. It of course has nothing to do with either Jerusalem or artichokes. Samuel Champlain observed them cultivated in the area of Massachusetts in 1605, and wrote that the flavor resembled that of an artichoke, which seems to account for that part of the name. A common explanation for the first part of the name comes from its membership in the sunflower family, the girasole, which is suggested to have been modified by folk etymology into "Jerusalem." However, this explanation has been demonstrated to be inadequate because the name "Jerusalem artichoke" had been established before "girasole" had been for the sunflower. A conjecture that matches the historic facts relates the name to Terneuzen, the place where they were first introduced into Europe. For a brief time the plant was known in Europe as *topinambour.* In 1613 six Tupinamba Indians from Brazil were exhibited in France and caused something of a sensation; Paris street vendors selling roasted Jerusalem artichoke tubers appropriated the name to add glamour to their wares. Modern vendors are trying to invent a new name to replace what they apparently feel repels trade by its ambiguity. For the Jerusalem artichoke, then, they have coined the name "sun choke." I question the wisdom of attempting to replace an ambiguous name with a silly one. Historically such attempts to gerrymander words have had a very poor track record for success.

As suggested by the fact that Jerusalem artichokes reseeded themselves with abandon in ancient Indian garbage heaps, they present few problems in cultivation. As with potatoes, a section of tuber having a bud or eye is used for "seed." According to one description I read, a pound should yield about twenty-five seed pieces. According to my experience that estimate is too generous. At any rate the globule of tubers is cut apart into however many sets it will provide. Plant them in the early spring, setting them about four inches deep and about eighteen inches apart. They grow well in a variety of soils, and produce vigorously with a minimum of attention. They put up a slender stalk, eventually reaching a height of six to eight feet, which, with the plant's leaves, distinctly shows its relationship to the sunflower. In late summer they produce small blossoms resembling a cross between a daisy and a sunflower. The blossoms are said to have a smell resembling chocolate, but it's a resemblance I have never been able to discover. The tubers may be harvested anytime after late summer. They may be stored like potatoes, but commonly they are left in the ground to be harvested as wanted.

Once established, a planting of Jerusalem artichokes tends to be rather persistent. It is hard to dig a hill without leaving enough small pieces to reproduce the following year. Therefore it is a good idea to plant them where a more or less permanent stand will not be unhandy. I have a planting between a property fence and a high road embankment, where its height is welcome and it is unlikely ever to become a nuisance. I have seen wild native plots where the plants had reproduced themselves so plentifully that by midsummer they were practically impenetrable.

Plant collectors have been effective in finding local varieties with superior qualities, particularly larger size. Quite a number of different varieties may be found in seed exchange listings, and in a few catalogues oriented toward collectors, although the variety usually offered by most commercial houses is the common small variety. I planted sets obtained from *Johnny's Select Seeds* which were described as having been collected from an old Indian site in Canada; the tubers are at least twice as large as those seen in produce markets, making them much easier to deal with in every way. A very few commercial houses offer varieties described as having been collected from ancient sites and also as being larger than those usually seen.

* * *

There is more history, romance and mystery surrounding the rude and homely potato than around the most swishy and showy tropical fruit or the darkest and most devious herb. I think of the potato in terms similar to those in which I remember a very elderly and gentle Indian who farmed down the road from us when I was a child, and who, I discovered years after his death, had fought beside Chief Joseph during the Nez Percé uprising. Like him, the quiet potato has a past that it does not easily reveal.

Modern genetic research tends to place the primordial potato in the high Peruvian Andes, perhaps in the region of the fabled Lake Titicaca, where the tubers are still grown as a staple. By the time of Columbus it had been disseminated through much of South and Central America, and a wide variety of locally adapted cultivars had been developed. After Columbus' second voyage it was introduced by him into Spain, where it was cultivated particularly because of its keeping qualities and nutritional advantages as a naval store.

As food for sailors it recrossed the Atlantic with Ponce de Léon to Florida, where it was "naturalized," essentially in its homeland. From Florida its track becomes faint, but it appears to have been taken to Virginia, perhaps by English privateers. At any rate it is next found growing in the New World in the Virginia Colony.

Again the track becomes faint. A revered tradition has it that it now leapfrogged back to Europe through the efforts of Sir Walter Raleigh, that Elizabethan man of all seasons who is credited with birthing the tobacco industry. He is supposed to have made the first planting of potatoes in the British Isles on his own large Irish holdings near Cork. Although both soil and climate were ideal for its cultivation, this planting, if it took place, was not a success. Something that grew underground in brown lumps that took the place of wheat was an idea whose time had not come, and the weaving of Ireland's and the potato's destinies was postponed for nearly another two centuries.

A different tradition says that Raleigh gave potatoes to Queen Elizabeth, who may have been half in love with her gallant admiral. She is said to have had them planted on the banks of the Thames by the royal gardeners. In midsummer this exotic vegetable from the New World was featured at a state dinner. Unfortunately Raleigh had neglected to tell the Queen that the part of the potato to be eaten was an underground tuber, and so it was the tops that were cooked and served for dinner. Associated with members of the nightshade family, which includes the potato, is a poisonous substance called sollenium that occurs in sunburned tubers as well as generally through the vines. How the Queen's cook prepared the vines does not survive, but reportedly all of the dinner guests were made thoroughly ill. According to the story, it was because of this disastrous dinner that in England the potato fell into bad repute from which it took centuries to recover.

Still another story asserts that what Raleigh initially introduced to the British Isles was not the potato at all, but the groundnut, *Apios americana,* a wild tuber of damp and marshy areas of the North American coast. In *The Tempest* Shakespeare alludes to an otherwise unidentified New World edible as a ground nut. This is sometimes identified as the peanut, but historically it does not seem likely that Shakespeare would have known the peanut. Most likely Shakespeare did indeed mean the groundnut, the *Apios americana,* although there is an outside possibility he could have meant the potato.

Potatoes became generally popular across much of the European Continent before they began to be grown in the British Isles. They seem to have finally found their niche in Ireland only in the eighteenth century, when they were apparently introduced, or reintroduced, from France.

Although only a relatively small number of different varieties of potatoes are found in commerce or are offered as seed for gardeners, internationally an enormous number of varieties exist. In the potato's South American homeland, there are varieties adapted to virtually all of the inhabited elevations of the Andes. A single farmer may grow several different varieties each year, all, of course, heirloom varieties. Recent attempts to persuade Peruvian farmers in the mountains to grow "improved" modern varieties for reasons of superior production, etc., have met nearly uniform resistance. Those farmers who have adopted modern varieties seem to grow them exclusively for export; for home consumption and domestic trade they continue to prefer and grow their old traditional varieties. Most guerrilla gardeners would wholeheartedly endorse both their attitude and their practice.

Although potatoes are propagated vegetatively by cuttings, they do also set real seeds. These seeds, produced in a pod that resembles a small green tomato or a large green nightshade berry, seldom "throw true" to their parentage. Consequently it is possible experimentally to produce innumerable varieties that are genetically unique, or nearly so. The likelihood of a given seedling having any particularly superior or desirable qualities is slight. However, as has been noted respecting the potato in Sweden, the possibilities of a diversifying gene pool are great.

Among the guerrilla gardeners there are several who are concerned with preserving and propagating varieties of potatoes. A Colorado collector maintains about two hundred varieties. Another has a collection of over five hundred. Numerous individuals maintain smaller numbers, and many unique varieties are available through the exchanges.

The vegetable that has been most debased by commerce is no doubt the tomato. Oddly enough, it was never as much of an item in its homeland as it turned out to be after it was taken to Europe. It originated probably in southern Central America, and cultivation of it was distributed primarily to the north, without its ever becoming a particularly important food crop. A cousin, the tomatillo, also known as the husk tomato, was considerably more widely grown and used. Its importance in Latin America continues today.

Exactly what happened after it was introduced to the Old World is uncertain, but it enjoyed a spotty and uneven reputation. One name applied to it was "wolf peach," which is still reflected in its proper botanical name, *Lycopersicon esculentum.* It also acquired the name of *pomme d'amour,* or love apple. It is speculated that this name may have originated as a corruption of *pomme d'or,* or golden apple. It is known that some of the tomatos in early cultivation were yellow in color. Another speculation is that the French name may have come from a corruption of *pomo dei Mori,* or Moors' apple, from the possibility that early specimens may have been obtained in Morocco by French sailors, who supposed they originated there (as happened, for instance, in the name of the New World fowl which continues to be known as a turkey). The reputation the tomato briefly enjoyed as an aphrodisiac, then, may be strictly the result of a linguistic accident.

Where the tomato's reputation for being poisonous first originated is uncertain. It may have arisen, as some have maintained, purely because it belongs to the nightshade family, some of whose members have very poisonous berries. It is also speculated that the reputation came simply as an extension of its supposed effects as an aphrodisiac. Certainly the Pilgrims of northern America, who held the tomato to be an abomination, would have been more terrified of an aphrodisiac than of a deadly poison.

This sinister reputation, however, was by no means uniform. The tomato seems to have been integrated quickly into use in Spain, and there seems to be no indication that the Italians showed the slightest hesitation in adopting it. In fact, the Italians were leaders not only in its culinary applications but in its breeding as well. Its acceptance in France was slow, and slower still in England. Very likely the English climate, which is rather unsuitable for its cultivation, did little to encourage the popularity of the fruit. When it was reintroduced into North America, it seems to have been largely as an ornamental. A story retold too often chronicles how one Robert Gibbon Johnson announced that he would publicly eat a tomato in Salem, New Jersey, one summer day in 1820. According to the legend, while doctors argued, preachers prayed and women fainted, he ate tomato after tomato. The legend states that this opened the floodgates, and tomato eaters have never had their complete fill since.

Probably if he were to repeat his demonstration with a modern supermarket tomato, Mr. Johnson might justifiably have second thoughts. The heirloom tomatoes grown and exchanged today by the seed savers might well be older than that hoary story. As described in their lists, some of these varieties are in the most restricted sense of the word heirlooms, having been grown by the gardener's family for generations. There is virtually no point at all in discussing a minimal list or even a test list of the varieties available. For one thing, a specimen should be tested against one's local conditions to accurately judge its value in respect to others. And of course the varieties are simply too numerous to represent with accuracy.

In 1982 I grew nineteen different varieties of open-pollinated tomatoes designated to be heirlooms. Because my growing situation is a difficult one, the results of my experience would be virtually meaningless to the majority of gardeners. All of these tomato varieties have been given high marks by a number of different authorities whose opinions I respect, yet of these nineteen varieties only one performed well for me. None of the other eighteen ripened fruit. This information is mostly meaningless to anyone but me, however, because an exceptionally cool spring and early summer held back growth and prevented blossom set. And early cold weather and frost held back ripening. The one tomato that performed well is called Bonny best, also known as parson's, or parson's pride. This is an old variety that was commonly grown commercially at least until the 1940s, and one that I always grow no matter what else I plant, because it always produces for me and I can't imagine a tomato being any better. This evaluation must be qualified by admitting that the Bonny best is for me a memory variety; my family always grew it when I was a child. Every gardener now has available such a smorgasbord of open-pollinated varieties of tomato to choose from that virtual freedom from commercial sources is assured. That this possibility exists is largely due to the expanding exchanges of the guerrilla gardeners.

A relative of the tomato that was once quite popularly grown but little seen today is the ground cherry. This is a distinct species of the family, and it is often confused with the tomatillo. Both have fruits contained inside an angular Chinese-lantern-like paper husk. In England the ground cherry is frequently called the Cape gooseberry, because it was popularly grown in South Africa, near the Cape of Good Hope.

While, like the tomatoes, they are cold sensitive and like hot weather, ground cherries are easy to grow. The small seeds may be started indoors for spring transplanting. I have always had good success putting the seeds in early May directly into the soil where they are to be grown. Unlike tomatoes, there is no particular urgency to get an early crop, so there is little real advantage to early starting. They grow into sturdy, compact and upright plants about two feet tall. Usually they set fruit prolifically, which appear first as tiny green "lanterns" and later, as they mature, developing into the characteristic paper wrapper. The fruit does not begin to grow until the paper jacket is almost full-sized. At maturity the fruit, about the size of a very small cherry, begins to turn a canary yellow. The fruit falls off and does not complete ripening until it has lain on the ground some time. Therefore one does not pick ground cherries, one picks them up. They will keep a surprisingly long time lying under their bushes, actually very late into the fall. The flavor of the fruit is mild and sweet, and they are especially attractive to children, who enjoy picking them up and eating them on a bright autumn afternoon. Adults tend to lose the knack of relishing them as much as children do. Ground cherries are traditionally used in a kind of preserve, which is a pretty yellow color and has a pleasant, mild flavor. If given a chance, ground cherries will readily reseed themselves from year to year.

The origin of peanuts has long been somewhat obscure or confused. Because they so quickly gained acceptance in Africa without the time or means of introduction being known, it was long thought that peanuts originated there, or that at least a species was native to that region. It is now certain that they originated in the New World, perhaps in the Caribbean. How they reached Africa is unknown, but they did so at an early date, and were soon widely disseminated. They seem to have been reintroduced into America from Africa in the course of the slave trade, along with their Swahili name, *goober.* Regionally the peanut continues to be known as a goober, or goober pea. There is some interest in the peanut among the seed exchangers. Only some five or so varieties are commonly grown commercially or seen in commercial catalogues. A great many more are preserved, and to someone whose growing conditions allow peanut cultivation, these alternate varieties should certainly be of interest.

By an odd linguistic twist, the Inca name for the sweet potato, *batata,* became applied to the white potato, to which it is not at all related. The sweet potato is a member of the morning-glory family, while the potato is a member of the nightshade family. The origin of the sweet potato is unknown, and no wild ancestor has ever been discovered. In the earliest contact of the Europeans with the Maoris of New Zealand, those native people were found to be growing a variety of sweet potato for which they too used the name *batata.* This evidence led to a belief that the plant had been introduced into the Americas by early European explorers, but that theory is no longer widely held. There is still no uniform theory explaining how the Maoris came by the vegetable and its name.

There has long been confusion between the sweet potato and the yam, and in America at least there is a tendency to use the two names almost interchangeably. In some regions the word "yam" is sometimes used exclusively for this vegetable. Technically, however, the yam and the sweet potato are different plants entirely, and completely unrelated. The yam is the root of a tropical plant, and the name is actually applied to several different tropical perennials having starchy roots. They sometimes grow to great size, and they are used in some of the same ways as potatoes and sweet potatoes. Yams, using the name properly, are not grown in the Americas. Often produce markets will offer both what they call yams and what they call sweet potatoes, the "yams" commonly being more bulbous, and red with yellow flesh, those they call sweet potatoes being more slender, with white or tan skin and a white or pale flesh. Whatever they are called in the produce market, they are all sweet potatoes.

There are, however, a great many different varieties of sweet potatoes cultivated. Many commercial catalogues will offer one or two, sometimes offering varieties adapted to northern conditions and short growing seasons. Catalogues directed toward southern gardeners offer a large to very large selection of different varieties. At least two catalogues are devoted exclusively to sweet potatoes. In addition, there are large numbers of varieties grown and passed about among the seed exchangers.

One of the spices Columbus was hoping to bring back if he discovered a direct sea linkage with India was pepper. He did not find pepper, but he did discover peppers. They are another New World vegetable contribution and of course are in no way related to the pepper tree. Many pre-Columbian peppers continue to be grown. Some of these old varieties can be found in certain commercial catalogues, but many more are propagated and collected by the gardening underground. I have tried growing three varieties, with results compromised by a poor growing season. I had no luck at all with a variety called Colorado, which is said to grow long red pods and is to be used dried. It is also said to be very hot. With mulato I had better luck. The pods are long and green, and it is used fresh. The fruit is attractive and quite hot. I used it as seasoning in chili, and it was excellent. The chili ancho also did well. The pods are about two inches long, thick and cone-shaped. They are also used green, are not especially hot, and have a pleasant flavor. I think they might be good pickled, although I did not have enough to try doing that.

* * *

There are a number of pre-Columbian squashes that are cultivated and exchanged among plant collectors. A few are also offered by some commercial houses. Space requirements have limited me to trying only one variety, the Guatemala blue. The vine grew vigorously and seemed to be extraordinarily drought resistant. One vine set several squash, but time ran out and only one matured. It was cylindrical in shape, about fourteen inches long and about three in diameter. The flesh was firm rather than crisp, and the layer of flesh slightly thinner than what one would expect in a comparable sort of commercial variety. We baked it in its shell and found it pleasant, but probably not as sweet as some baking squash.

In contrast to the New World varieties, many of the Old World vegetables are of surprisingly recent origin. Cabbage, lettuce, peas, fava beans, lentils and most of the small grains, to name a few, are quite ancient. However, turnips, broccoli, cauliflower, brussels sprouts and rutabagas, again to name only a few, were relatively late in development or in dispersal, many dating after the discovery of America. Turnips, for example, were not introduced into England until the seventeenth century, at which time they caused something of a revolution in the livestock industry because of their use as feed for cattle. Seen in this context, it becomes easy to understand the quick acceptance of the New World crops when they appeared not in competition with similar foods but as a whole new dietary concept. It has often been observed how vastly more valuable to the Old World were the New World's foods than its precious metals. Very likely this difference in importance was more quickly perceived by the European common man than we might today suppose. Food affected his daily life almost immediately, the precious metals never.

When looking for heirloom vegetables among the Old World types, then, one is not necessarily looking particularly far back into the past at all. Even such ancient vegetables as carrots and cabbage were likely in their earlier forms coarser and less delicate than later cultivars. With Old World vegetables as well as New World introductions the heirloom gardeners and seed exchangers have numerous options for propagating varieties not usually available through commercial sources. Many of these include varieties the Common Market restrictions are taking off the market in Europe, and without the efforts of the heirloom gardeners these could well become extinct. Here again, for the sake of illustration, only a few varieties need be mentioned.

Of several varieties of European lettuce I have tried, two have excelled over every similar lettuce I know. A cos lettuce from England, called Lobjoits green, is distinctly superior to the varieties of romaine commonly offered by American commercial houses. As grown in my garden, the bunches were taller and denser, and the leaves crisper and more flavorful than commercial American cos I was growing for comparison. It also seemed more resistant to bolting, although I have not yet had enough experience with it to make an accurate judgment.

The other variety is clearly the best lettuce I have ever had. It is called Little Gem and has been described as *the* gourmet French lettuce. It is similar to the so-called bib lettuces, although slightly more upright in growth. It is the only lettuce I have ever eaten for which the word "buttery" was truly appropriate. It was universally praised by all who tried it. When hot weather came, however, it was one of the quickest to bolt. A fall planting did very well, remaining good until ruined by heavy freezing.

A salad vegetable which appears to be little known in this country and which where known is usually considered an object of gourmet eating is called "corn salad" (*Valerianella locusta*). My seed came from Holland. It is a low-growing plant, resembling in habit and even somewhat in leaf shape, juvenile mustard seedlings. For me it was not very productive, and when the weather turned warm it wilted unless watered almost daily. The flavor, however, is fine: lettucelike perhaps, but unique. Mine did not flower or produce seed. Its cultivation in American gardens needs to be better understood, for it is something many people might be interested in growing.

Two French cucumbers I have tried were both pickling varieties. One, called the Parisian pickling, was quite productive, and the main part of the crop matured close together. It was good as a table vegetable, and did indeed make excellent pickles. I made a crock of slow brine dill pickles which came out firm and crisp. They were fine for pickling, but I'm not sure I would rate them particularly above cucumbers more commonly found. Others who are more skilled and demanding in the making of pickles might disagree.

Another French variety, cornichon de Bourbonne, is grown to make a tiny sour pickle. However, I could not locate a recipe and had to pass up that opportunity. Allowed to grow large, they were a good table vegetable. My plants were not especially productive, and if all of the cucumbers were to have been pickled small, my planting would not have provided a particularly impressive batch.

Of the different European varieties of carrots I have grown, most were superior, or conspicuously superior, to the usual commercial catalogue offerings. The single exception was a variety called nugget, which is said to be especially suitable for heavy soils. The mature carrot is short, about two inches long, and shaped like a wide, blunt cone; it makes for an odd-looking carrot. They grew fine, but their flavor was not particularly distinguished, being rather overpoweringly carroty.

The French primenantes variety of carrot is especially sweet and crisp. The carrots have an attractive bright-reddish color, and mine grew large, retaining a smooth cylindrical shape to maturity. Sucram carrots, which are reputed to be generally popular in Europe, are used basically as a miniature or a so-called carrot tip. They never grow to the expected size of a carrot and are obviously not a carrot to be used for storage. They make an attractive pack as canned carrots. Like immature or baby carrots, they have an especially sweet flavor. My children cleaned them out before I could see how large they would actually get if left the full summer in the ground.

A French variety called touchon is another fine option. Mine grew long and straight and were medium to large in overall size. The flavor is particularly sweet and good. They are said to be an especially good carrot for juicing. They also make excellent carrot sticks and fodder for children.

One Oriental variety called the Imperial long, from Japan, is occasionally seen offered as a novelty item. It is said to grow up to three feet long. Mine did not grow for a long time, but they did grow very long, slender, and remarkably smooth. There was nothing to particularly recommend in their eating quality. A large number of mine, over half, went to seed their first year. I don't know if that's characteristic of the variety, but it is hardly a commendable quality.

Like most of the melons, cantaloupes originated in the Near East. They may have been brought back to Europe by the Crusaders. At least one place where they showed up in the Middle Ages was Italy, where they were grown on the estates of a castle near Rome called Cantalupo. The name means "wolf howl" and accounts for the name of the subgroup, *cantalupensis*.

Although cantaloupes have been adapted to widely different growing conditions than those of their hot desert homeland, their inclination is still to prefer a deep sandy soil and lots of summer heat. The breeding efforts that extended their range have at the same time reduced their virtues. Commercial cantaloupes have been as badly debased as commercial tomatoes, and for about the same reasons. A ripe cantaloupe can no more be shipped profitably than a ripe tomato. Since the last thing the leaves put into a cantaloupe is sugar, when they are picked they are as sweet as they are ever going to be. Although they will soften and gain color in storage and shipment, they will not sweeten beyond the level they had when picked. As with tomatoes, therefore, anyone who desires good cantaloupes must grow his own. But as is true for tomatoes, for optimum quality he must go beyond the commercial varieties.

Fortunately there are many excellent noncommercial cantaloupes to choose from, both from the collections of seed exchangers and from a limited number of catalogues specializing in foreign seeds and gourmet varieties. In 1982 I planted two French varieties and one Oriental variety. The French cantaloupes, both given highest commendations, were Charentais and Chaca hybrid. The Charentais is large, sweet and, at ninety days, reasonably early for a large, good-quality cantaloupe. The Chaca is small and, at sixty-eight days, very early. It is reputed to be especially sweet. Both grew well for me and produced heavily. The honey gold is an Oriental variety which can be eaten rind and all. It is particularly adaptable to trellis culture. Mine were not particularly productive and did not ripen to their potential. They were good, but I knew they would have liked to be better.

Whatever varieties of vegetable a gardener grows, it is almost certain that there are alternative varieties being grown by the guerrilla gardeners that are superior. Coming into contact with this underground community is almost like discovering a second New World with another cornucopia of new treats for a strangling Old World diet of stale and tasteless crops. Especially with temperamental vegetables, there are likely to be somewhere among this brotherhood varieties which have been adapted to almost any conceivable special problems. I may even find a cantaloupe I like that will actually ripen properly in my climate.

Planting Guide

This suggests in terms of relative hardiness the appropriate times for planting or setting out vegetables for optimum production and efficiency of crop rotation.

Two months before last frost	One month before last frost	Last frost	Two months before first frost
VERY HARDY	HARDY	TENDER	HARDY
Broccoli	Beets	Corn†	Beets
Cabbage	Carrots	Tomatoes	Collards
Lettuce	Chard	Eggplants	Kale
Onions	Mustard	Peppers	Lettuce
Peas	Parsnips	Sweet potatoes	Spinach
Potatoes*	Radishes	Beans	Turnips
Spinach		Cucumbers‡	Onion seed§
Turnips		Okra	Broccoli§
Kale		Soybeans	
		Squash	
		Melons‡	

* Folk wisdom has it that potatoes should be planted on Good Friday.
† Countrymen say corn should be planted when the oak leaves are the size of squirrels' ears.
‡ Melons and cucumbers can be given an earlier start if planted under the protection of paper caps, etc.
§ Onions and broccoli are commonly planted in the late summer for early crops the following year.

3. THE PRACTICAL SEED SAVERS

Any gardener who develops an interest in heirloom plants or other unusual varieties will probably wish to save seeds from his own plants for his future use as well as for trading and sharing. If he wishes to become involved in exchanging seeds with other gardeners, saving seeds will certainly be a necessity. There is no particular mystery about growing and saving one's own seeds, and it is really not difficult to save them from year to year. If it is done on anything like a large scale, however, a few tips can make it easier to keep strains pure as well as to develop one's own private strains.

One important biological concept is essential for seed saving. Many of the varieties of seeds commonly offered by seed houses are hybrids. This means that a special selective sequence of plant breeding results in a variety that has desirable characteristics, *for one generation.* The offspring of this desirable plant, however, will not have the particular qualities of its parents. In fact, it may have undesirable characteristics or even be, for all practical purposes, useless. To produce the sought-after characteristics, the exact breeding sequence must be repeated for each generation. Therefore it is pointless for a gardener to collect seeds from hybrid varieties to propagate another year, because they will not "throw true" to type. Unless one is a plant breeder capable of reproducing the crosspollinization sequence necessary to produce a particular hybrid variety, one must buy seed for each planting from a commercial supplier who is capable of making the necessary crosses.

Many tomatoes, most swee. corn and field corn, many

melons, etc., are hybrids, and they will be advertised as hybrids in the catalogues or on the seed packets. Plants which are members of a "true" genetic strain and not the result of hybrid cross-pollinization will "throw true" to type from seed generation to generation. These plants are called "open-pollinated" because they pollinize themselves (usually with the help of wind or insects) and do not require being pollinized by hand, with the plant breeder selecting which strains will pollinize which to produce the desired end result. Most of the varieties which interest the plant collectors, the seed savers or the guerrilla-gardening fraternity are generally open-pollinized.

Most fruits, however, even the very ancient strains, will not throw true from seed and must always be reproduced by vegetative reproduction such as grafting, rooting cuttings, layering, dividing tubers and corms, etc. Plants reproduced in this way are called "clones." In spite of the fact that this is a trendy term with science-fiction-like connotations, the practice is very ancient. If one plants a graft of the cherry known as the Spanish sweet cherry, for example, he does not plant the *same kind* of cherry grown in Pliny's orchard in ancient Rome, he plants the *same tree.* There has been no germ exchange through sexual propagation in the Spanish sweet cherry since Pliny described the fruit; it has been reproduced strictly through cell division derived ultimately from the original plant. Cloning.

Such thoughts add a special romance to growing varieties of heirloom fruits. One can not only grow Pliny's ancient cherry, one can grow the greengage plum Sir Thomas Gage introduced to England from the French court so many centuries ago, and the *court pendu plot* apple introduced to Britain during the Roman occupation. What a great deal more panache there is to grow such venerable fruits than a variety introduced, for example, in 1976 by the experiment station of a local university.

If one wishes to save seeds from open-pollinized plants, a little basic attention to plant selection is required of the gardener. A few decisions, conscious or even partially unconcious, need to be made in deciding which plants should be chosen for their seeds. Basically, one will want to save seeds from plants that are particularly good-tasting, especially early, notably hardy or simply attractive. One may save seeds from a particular pepper or tomato, for example, for the most simple of all reasons, that that particular specimen was the only one in the patch that matured fruit. Offspring of this plant have a better chance of producing mature crops than unselected varieties; the offspring of those crops have still better chances, and so on. In this way one may produce varieties that are tailored to one's own growing conditions. Or, one eats a cantaloupe or watermelon and is so delighted by its flavor that he cannot help saving its seeds and thus is on the way to tailoring a variety to his own taste preferences.

More systematically, a gardener may select from among several tomatoes or cantaloupes plants that matured the first fruit, or the ones which produced fruit with the best flavor, fruit that most resisted cracking, rotting, wilting, insects, etc. In time, actually in a relatively short time, this will result in a subspecies of a particular plant that is specifically suited to one's own garden: one's particular soil, climate, even taste preferences. On the farm where I grew up, my father began growing rutabaga seeds for his commercial planting because he was dissatisfied with the quality of the rutabagas grown from the seed he was getting from the seed companies. By his selecting for seed stock those roots with qualities he liked, rather dramatic improvement in crop quality was apparent in only two or three years. Soon other growers were coming to him to buy his seeds. My father did the same kind of selection with carrot and onion seeds, both of which he grew as field crops, and with similar although not as dramatic results. By selection, he produced local seeds that grew much better crops under local conditions. Each gardener can do the same for himself by watching his crops and saving the seeds of plants which have qualities that he wants. He may also "shop" in the gardens of other people, or occasionally even in the wild, for plants with qualities he would like in his own plants.

There is an old saying that trees should be moved north of

their native range or habitation in increments of no more than forty miles. Thus, theoretically, if a nut tree is moved no more than forty miles north of a place where it is hardy, it will adapt; and its seed moved no more than another forty miles north will adapt, and so on. There may be nothing but folk wisdom in this particular formula, and to what truth there may be in it there are obvious limits, but there is a basis of truth in the principle nevertheless. Moving by short distances at a time has enabled many species to be grown far outside the climate of their origin.

Corn is a prime example of a crop moved far north of its original range. Corn appeared first in prehistoric times in the tropics, near the equator, but it now grows in the extremes of the North Temperate Zone. Although it has adapted to those northern ranges, almost whimsically, corn doggedly retains one of its tropical mannerisms, that of preferring to pollinize when days and nights are approximately equal in length, as of course they always are at the equator. Typically corn requires a day length of no more than twelve to fourteen hours to produce tassels. This effect is called a "photoperiodic response," and how it works is not yet fully understood. Most gardeners at one time or another have attempted to get early corn by cunningly managing by whatever means, including luck, to get large corn plants at the very beginning of summer, only to have them dawdle along and end up not producing much earlier than his regular planting. The culprit, and the limiting factor, is the corn's "memory" of its place of origin. Corn grown in the more southern parts of the United States set earlier in the summer because in these regions there is never the dramatic shift in day length that there is in the more northern parts of the country. A comparable kind of response to day length can be noticed in dandelions, which, although they bloom more or less continuously from spring through fall, have a major flush of blooming near the vernal and autumnal equinoxes, when days and nights are most closely equal in length. The farther north one goes, the more strikingly this effect is seen.

Corn can also be used to illustrate another principle a gardener should take into account in selecting parents for seed saving. Put simply, a gardener should fully and consciously realize what qualities he is truly selecting for. In the nineteenth and early twentieth centuries "corn fairs" were popular in rural America. At these fairs corn was judged by the qualities of individual ears, which led to the development of handsomer and handsomer ears over the course of years of selection. However, in the first decade of this century studies of corn which concentrated on the yield per acre rather than on the appearance of individual ears began to be made. The results of these studies showed that "specimen ears" performed very poorly if judged in terms of yield per acre. From these studies began the breeding programs that have led, for better or worse, to the extremely high per-acre yield of modern corn. In other words, a gardener should select seeds toward ends he really wants, not ends he thinks he wants. Otherwise he might end up rediscovering the square machine-picked tomato.

For most crops it is preferable for the gardener to select seed from several different plants in his garden rather than from a single specimen. If only one plant is used for each year's seeds, the effect will be over the course of years to narrow and limit the gene pool, or inbreed the crop he is propagating. Inbred plants may lose vigor, quality, resistance to disease and insects, and fertility. Pumpkins and squash, however, seem not to be subject to inbreeding, so there is no reason not to select all of one's seeds for these plants from a single fruit.

If a gardener's needs or wants are exacting, probably he will want to establish a file on each strain he is perpetuating or improving. Many qualities, for example rate of germination, resistance to wilt, vigor of early growth, etc., may not remain visible when it is time to harvest the seeds, and must be noted as they are observed, while the plant is growing and developing. If an extensive number of different seeds are being perpetuated, it is always advisable to establish some system of records. Not meaning to put a morbid face on the matter, such information would be useful indeed to one's heirs.

A gardener who is perpetuating species by saving seeds should know a few basic things about the physiology of plants. Not all flowers are alike in the way they fertilize and are fertilized and in the way they bear fruit. There are basically three kinds of blossoms: some blossoms are *perfect,* meaning that each flower has both male and female parts; some plants bear two kinds of flowers, male and female, and are called *monoecious;* some plants bear male or female flowers, but not both, and are called *dioecious.*

Some plants have flowers that exclusively, or almost exclusively, fertilize themselves, and are said to be "self-pollinated." Only perfect flowers may self-pollinize. With some plants the pollen is carried from one flower to another in an exchange that is called "cross-pollination." The flowers of some plants bearing perfect flowers may cross-pollinate as well as self-pollinate. Cabbage, for example, has perfect flowers, and it cross-pollinates readily. Tomatoes have perfect flowers, but they cross-pollinate little because they are not especially attractive to insects. Beans have perfect flowers but

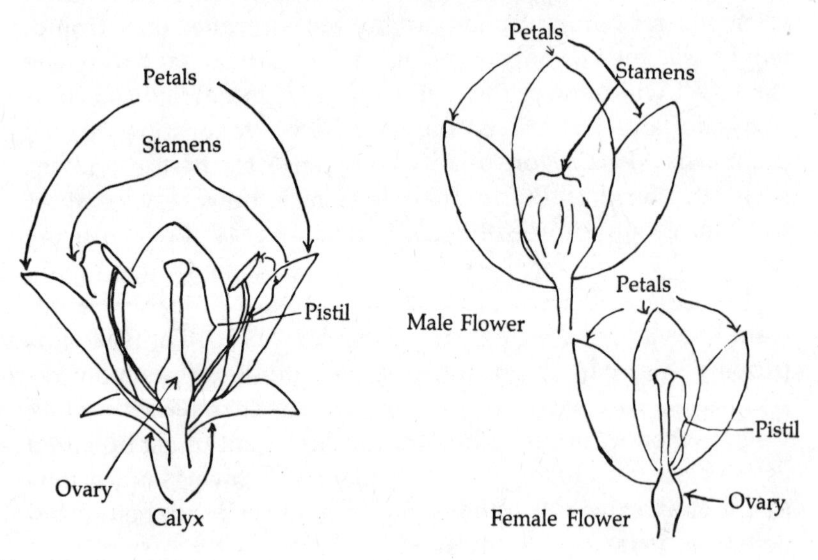

Perfect Flower

Monoecious Flowers

cross-pollinate little because fertilization usually occurs before the blossom opens.

Many garden plants are monoecious, bearing separate male and female blossoms on the same plant, as for example the squash and its relatives. Usually the first blooms which show are male, and many people commonly call these blossoms "pollinizers." These blossoms are easily distinguished by the fact that the bloom seems stuck on the end of a stem, with no bulge on the top of the stem where it joins the petals. Inside the blossom are pollen-bearing stamens, usually clustered or growing together as a single upright appendage in the center of the flower. In a large blossom this stamen or stamen cluster may be an inch or more long. The female blossom has an enlarged ovary at the top of the stem joining the base of the petals. This enlargement is a miniature of the fruit the ovary will eventually develop into. In such squash as the zuccini or the yellow crookneck the fruit is picked so soon that it is scarcely more than a fertilized ovary. The male blossom remains open a day or two, withers, and falls away. After the female blossom is fertilized, the ovary begins to swell, and, while the blossom withers, it usually remains attached to the end of the developing fruit, sometimes until the fruit is completely mature.

Plants which are dioecious will bear flowers that are either male or female, but never both. Such plants as spinach, asparagus and hops are dioecious. Since it is not necessary for the female blossom to be fertilized for hop production, commonly plants are reproduced through root cuttings, with only female plants being grown. It seems likely that female asparagus plants produce more and better spears, and there is some experimentation under way studying the possibilities of limiting plantings to females only. The holly is a familiar example of a dioecious tree. Only the female bears the attractive red berries, but without a male tree the flowers will not be pollinized and no berries can be grown; it is therefore necessary to have at least one male tree for every several females in order to have berries. The date palm is also dioecious, with one male tree being adequate for many females. It is common for growers to limit planting to female trees and to buy pollen for fertilization of their crop.

Because of differences in the structure of their flowers, some plants are much easier than others to propagate for seeds. All open-pollinated plants may be propagated by the home gardener, but some require care to sustain purity of strain. Among the easiest are beans and peas. Both have perfect flowers, with male and female parts. Because the pollen ripens and the pistil is capable of being fertilized before the blossom actually opens, seldom is there cross-pollinization between or among varieties of beans and peas. For the seed saver it is very convenient to have blossoms pollinized before opening, because this makes it much easier to grow a number of varieties at the same time and still keep each strain relatively pure. To be absolutely certain of purity, grow the bean or pea rows of different varieties in different parts of the garden, or with the rows separated by a relatively tall-growing crop.

To collect beans or peas for seeds, harvest when the pods have the brittleness of a dry leaf. The pods may be picked by hand, or the entire vine may be pulled up and taken under shelter to complete drying after the pods have matured, but before they will shell easily. When dry, the pods may be shelled by hand, or the entire dry vine may be put into a sack and flailed or manhandled until the beans or peas have all fallen out of their pods. With some varieties the dry pods pop open almost with a touch, but some grimly hang on to their seeds to the bitter end. Many varieties grown for green beans are quite hard to shell. Once the beans or peas are hulled, it is not difficult to separate the shelled seeds from the rest of the dry debris by winnowing.

All of the members of the Brassica family, which includes cabbage, kohlrabi, kale, brussels sprouts, cauliflower and broccoli, have perfect flowers and are cross-pollinized by bees and other insects. Consequently, if more than one member of this family, or more than one variety of a given species, flower at the same time relatively near each other, some degree of undesirable cross-pollinization is likely to occur. For absolute purity, no related members of the family should be blossoming within a quarter of a mile. Relative purity can be obtained if blossoming plants are at least one hundred feet from each other, or separated by such tall-growing crops as corn.

Members of the Brassica family tend not to be self-fertile; that is, the blossoms on a single plant do poorly in fertilizing each other. For proper pollinization, therefore, two or more plants of the variety which the gardener wishes to propagate should be grown near each other. As widely different as the edible parts of the members of this family are, their blossoms, pods and seeds are very similar. The fruit of cabbage, broccoli, cauliflower and brussels sprouts are all buds, and when they begin to flower, the flowering structures are very similar. Branching stems will emerge from the heads or buds, with yellow blossoms finally opening along the stem. The seeds are borne in slender pods an inch or two long. Since the pods ripen unevenly from the main stem outward and upward to the top of the seed stalks, the seed stalks ought to be harvested before the pods are fully ripe. I cut the seed stalks when most of the pods are fully formed and stuff them head first into a large paper bag to complete ripening in the sun. When brittle dry they are easily thrashed within the bag and cleaned by winnowing in a light wind.

Since the part of many members of the Brassica family that is eaten is actually a bud, plants must be allowed to go through the edible stage and complete the blossoming cycle in order to grow seeds. Most members of the family are biennial, at least in their northern ranges, producing buds the first year and blossoms and seeds the second. If winters are not too severe the plants may simply be left in the ground to winter over. With such protection as a covering of leaves or straw, the members of the family can survive quite hard winters. Otherwise the individual plants are dug, roots and all, and kept in cool storage with their roots protected from drying until early spring, when they are returned to the ground. For seed production, one will naturally select the specimens with the most desirable characteristics. Because the head of a cabbage plant is usually quite dense, after replanting make a shallow cut across the top of the head. This provides an easy exit for the seed stalk as it begins to develop. Otherwise it may emerge twisted or deformed, or perhaps not emerge at all. All members of the Brassica family are stimulated to blossom by the long day length of early summer, so it is important that they be planted as early in the spring as possible to become established and in the proper stage of development at blossoming time.

Broccoli is an exception to the pattern of the rest of the family. It is an annual, and produces seed in its first year. Therefore, for best seed production it is necessary for the plants to be given a very early start to insure that when the long days of early summer arrive they are far enough along to develop blossoms from their "heads." The only variety of the Brassica family that I grow for seed is a broccoli descended from seeds I obtained in England. Unlike our common broccoli, this is biennial or even perennial in habit, producing heads or buds in early spring or late winter of its second season. In my Zone 5 climate I plant seeds in late summer and get returns in the early spring. For seed I select for hardiness and earliness. I have never had a plant survive more than one winter, but progeny of this seed growing in a Zone 6 area has survived two and three years.

Chinese cabbage, despite its name, is not related to the Brassica family, and will not cross-pollinize with cabbage, etc. The plant is an annual, producing seed the first year. It has perfect flowers and is capable of cross-pollinizing with mustard as well as with its own varieties. To maintain the purity of a strain, it must be grown in isolation from other family members. The seeds are harvested in the same way as members of the cabbage family.

Despite superficial physical resemblances, lettuce is not a relative of the cabbages. An annual, lettuce produces seed the first year. The flower is perfect and almost always self-pollinizes, so several varieties may be grown near each other without crossing. However, my family always saved seed from leaf lettuce, and suddenly one year our lettuce became inedible: bitter, and with a very unpleasant flavor in general. It was decided that our seed had crossed with Chinese lettuce, a weed almost ubiquitous in our area. Whether that is indeed what had happened I don't know, but the lettuce had gone bad. However, in recent years I am again saving lettuce seed—romaine and English bib—and have had no problems, although there is also Chinese lettuce growing in my area.

When lettuce bolts, normally the gardener removes it and probably plants something else in the vacated area. If it is allowed to remain, the stalk which bolts from the center will produce a cluster of upward-growing stems with numerous small blossoms on the tips. These flowers eventually mature a cluster of tiny white fluffs resembling miniature dandelion parasols, each attached to a seed for wind-borne distribution. The seeds do not all mature at once, which slightly complicates harvesting, since the mature seeds tend to detach and float away in a strong wind. Some people harvest seed in stages, placing a large paper sack over the entire seed head when the first fluffs appear, then bending the stalk down and shaking it vigorously. The process is repeated over the maturing period. Or, and this is what I always do, wait until most of the blossom heads have matured but before the stage when the wind may detach them, cut the entire head off and insert it upside down in a paper bag. I leave them outside protected from wind and rain until the stalk is brittle dry. Vigorous shaking then dislodges the majority of the seeds. Because it is quite small and light, lettuce seed is difficult to get thoroughly clean. With a little patience, it can be gotten clean enough for home garden use.

Onion seed is rather easy to grow, but, because varieties cross-pollinize readily, for practical purposes only one variety can be grown at a time. Although onions normally are biennial, flowering the second year, in a given planting there will almost always be some that bloom the first year. Since this is an undesirable characteristic, such plants should never be used for seed stock. In areas where winters are not too cold, onions will often over-winter in the ground; with a protective covering they can be made to survive considerable cold. However, since one is saving seeds to perpetuate desirable characteristics, when onions are allowed to over-winter in the ground one should inspect them and replant only those that pass close scrutiny. If not replanted immediately they may be kept as any other winter onion, in a cool, dry and airy storage area. They should be reset in earliest spring. Each onion bulb will send up one or more stalks, each with a globular umbel or seed head on the top, consisting of many spikelets radiating from the center, with a blossom on the tip of each spikelet. As the seeds develop, they will be visible on the tips, BB-sized black spheres bursting through the dried flower. Cut the stalk when it seems that seeds are visible on about half of the spikelet. Set aside the seed heads attached to their stalks to dry in the sun or under protection from rain, with good air circulation. When thoroughly dry, the seeds may easily be rubbed out between the hands. The seeds themselves are somewhat fragile and may be broken if thrashed too vigorously with such tools as a flail, etc. Because the seeds are fairly large and heavy, it is not hard to clean them thoroughly by winnowing.

Spinach is a somewhat curious plant. It is an annual having both monoecious and perfect flowers. That is, some plants have male blossoms, some female, and some plants produce blossoms each of which combine male and female parts. The blossoms are essentially wind-pollinized, and the plants produce a very fine pollen that travels a great distance. For practical purposes only one variety can be raised for seed at a time. As for kitchen use, plant the spinach early in the spring. Normally one simply selects a few plants from his first planting for seed purposes. When the spinach begins to bolt and form blossoming stalks, select for their good qualities however many plants are needed. Since resistance to bolting is one of the desirable qualities selection is usually based on, make your choice from among those which bolt last. However, since the bolting of spinach is an effect of day length and not of temperature, there are limits to the control of bolting that can be achieved by selection. When all seed stalks have turned brown, cut them and lay them aside to complete drying, either in the sun or under cover with free ventilation. When dried, they may be thrashed by hand, and the seed cleaned by winnowing.

Probably more gardeners are interested in saving and perpetuating tomato seeds than seeds of any other garden crop. The tomato has a perfect blossom and usually is self-pollinated, making it fairly easy to maintain a number of different strains simultaneously. It is possible that there may be a slight amount of crossing, so for absolute purity different varieties should be separated by a short distance. Bees show little interest in tomato blossoms, which makes it unlikely that pollen will be transmitted any considerable distance. As is true in the case of the majority of vegetables, most gardeners who collect seeds save them from plants they are growing to eat. Plants that show particularly good qualities may be marked for attention: those most true to type, those earliest to set fruit, those earliest to ripen, those with the most-thrifty vines, with the best-quality fruit, etc. Select fruit growing from more than one plant to provide genetic diversity and prevent eventual inbreeding. Naturally the individual plants selected should each conform to the ideals or standards of the individual gardener.

To harvest seed, pick individual tomatoes when they are dead ripe, and crush them together in a bowl or pan. Allow the pulp, seeds and juice to ferment at room temperature, stirring once or twice a day. The process of fermentation will take three or four days. To clean the seeds, add more water and allow the seeds to settle out. Drain the scum and the liquid, then add more water again. It will take three or four washings to fully separate the seeds from the tissue and fibers. Seeds that float to the surface should be discarded with the rest of the debris, because they will be of doubtful fertility. The purpose of the fermentation process is both to aid in cleaning and to kill certain disease-causing microorganisms. When the seeds are fully cleaned by washing, the handiest way to dry them, since they are quite small, is by spreading them on sheets of paper. Spread them very thin. When they are dry, separate those seeds which have stuck together by rubbing between the hands.

Peppers, like their cousins the tomatoes, have perfect flowers and are mostly self-pollinated. As with tomatoes, for highest purity segregate the plants of different varieties a short distance apart, or interplant between rows with another vegetable. Like tomatoes, peppers should be allowed to ripen on the vines before harvest. Open each pepper by cutting lengthwise, strip out the seeds, and allow them to dry on sheets of paper or toweling in the open air. They may be washed before drying, but need not be fermented. When they are thoroughly dry, it is not difficult to finish cleaning the seeds by rubbing.

Cucumbers are monoecious, having separate male and female blossoms on each individual plant. Cucumbers will not cross with other vining crops—melons, squash, etc.—but varieties will cross with each other, and all varieties will cross with gerkins. Practically speaking, to grow cucumbers for seed, only one variety may be planted at a time. If one has adequate space, relative purity may be achieved by separating varieties by at least one hundred feet; since they are pollinized by bees, complete purity requires separation of varieties by about a quarter of a mile. To save seeds, select and mark preferred vines, and select preferred fruit according to size, shape, time of maturity, sequence of maturity, productivity, etc. Since by the time seeds are collected the fruit is beyond the edible stage, taste testing will have to be conducted with other fruit from the same vine. To maintain genetic diversity, fruit should preferably be collected from more than a single vine. Wait until the fruit has ripened to a golden yellow before picking for seed. Open each cucumber and remove the seeds and the interior pulp and put them into a jar; mix this with water and allow the mixture to ferment about four days, stirring it once or twice a day. Pour off the liquid and the scum, then keep adding water and draining until the seeds are clean. Seeds that float to the surface should be discarded. Spread the seeds on paper to dry, either in the sun (protected from wind and rain) or inside with good ventilation. When they are thoroughly dry, cleaning the seeds may be completed by hand and by winnowing.

Cantaloupes will not cross with the other vine crops, but varieties will cross among themselves. The same cautions that apply to cucumbers should be observed here. It is difficult or impractical to grow more than one variety of cantaloupe at a time if seed is to be saved. When making selections of seed specimens the individual cantaloupe may be tasted, and seed saved only from the best-flavored specimens, although matters of hardiness, earliness, productivity, etc., should be taken into consideration as well when making seed selections. On my parents' farm, saving cantaloupe seed was a family affair. Since we had cantaloupes in great abundance, only the best-flavored were completely eaten; rejects went to the chickens. And only the seeds from cantaloupes of exceptional flavor were saved. The rare cantaloupe that was both called and chosen had its seeds deposited on sheets of newspaper laid away in a dry shed. When all of the cantaloupe seed was cleaned and put together at the end of the summer, there might be a quart or more of seeds, more than enough to plant our small truck-garden melon patch as well as to give to anyone who had not saved seeds himself. Unfortunately I have long since lost contact with that strain of seeds, a misfortune frequently seen regretfully repeated by correspondents in the Seed Savers' Exchange and elsewhere. Basically, however, that is the way cantaloupe seed is saved. Select a small quantity of seeds and pulp from the seed cavity of several different desirable melons. Place them on paper and allow them to dry. If the air is moist or humid, spread the seeds to prevent molding, and turn them occasionally as the drying progresses. When they are thoroughly dry they may easily be cleaned by hand and by winnowing.

Watermelons also are monoecious, being pollinized chiefly by bees. As with cantaloupes, the different varieties of watermelons cross-pollinize readily, and all will cross with citron melons. Therefore to save seeds from watermelons more than one variety may not be grown at a time unless they are separated by considerable distance. As with cantaloupes, seeds should be collected from more than one vine to maintain genetic diversity. Collecting seeds may also be done while eating, allowing the last and crucial test, that of flavor, to be determined on the spot. It is also possible to pulp the inside of a watermelon, put the seeds and the mush into a jar with a goodly quantity of water, and by agitation separate the seeds from the residue. It is not necessary to ferment them. They may then be spread on paper to dry and finished by hand cleaning.

Saving seeds from the various squash plants gets a little more complicated. All belong to a common genus, *Cucurbita.* But to know which will and will not cross-pollinize it is necessary to place each variety in its proper species. Basically there are four species of squash, and if seed is being saved only one member of each species may be grown. The species *Cucurbita pepo* includes all of the summer squashes, pumpkins, acorn squashes, most of the gourds, and spaghetti squash. The stems of C. *pepo* are five-sided and have prickly spines, sometimes painfully prickly in the large coarse varieties. More than one variety of this species may be grown together for their fruit, because the fruit will not be affected by the cross-pollinization. However, if seed from cross-pollinized fruit is planted, the crop will probably bear some resemblance to each of its parents, but will most likely be inferior to both, or even worthless. In 1982 a seed from such a cross came up as a volunteer in a corner of my garden. Since it was in a corner and not particularly bothering anything, I let the little stranger grow all summer. From the appearance of the fruit, the seed seemed likely to have been a cross between a pumpkin and a zucchini, both of which I had grown the previous year. While the vine was prolific, except to satisfy my curiosity the fruit was of no value. Although they were large and yellow, they were too zucchini-shaped even to make jack-o'-lanterns out of.

The second species of the cucurbits is called *Cucurbita maxima* and is characterized by a stem which is soft and round and covered with hairlike growths rather than spines. This species includes the hubbards, the bananas, the buttercups, the mammoths and the turbans, plus other named varieties of the same general kind. Crosses among these members may not be so disastrous as to produce squash that are actually worthless, but crosses will wipe out the advantages specific varieties are cultivated for.

The *Cucurbita moschata* group have a stem that is five-sided but smooth rather than spiny or hairy, and the fruit stem enlarges and flares where it joins the squash. Members of this species include the butternut (as distinct from the buttercup), the cushaw and the Kentucky field, plus other named variants of these varieties.

The fourth species is the *Cucurbita mixta,* which is physically similar to, but genetically different from, *C. moschata,* with which it was formerly catalogued. The differences are quite subtle and require experience and training to determine. This small species includes the Tennessee sweet potato, white and green cushaws, and a few others.

A gardener can grow one of each of these species for seed with no trouble, but if more than one variety of each species is grown they will cross-pollinize. They will not, however, cross with any of the varieties of melon.

For seed saving, select ripe fruit (zucchini and other summer squash which is eaten immature must be allowed to remain on the vine until the seeds reach full size), separate the seeds from the flesh, and dry them on a sheet of paper. Particularly the largest varieties of seeds should be spread thinly and turned frequently while drying to prevent mold. When the seeds are thoroughly dry, finish cleaning them by hand.

Corn is one of the more difficult crops to maintain purely. It is monoecious, the conspicuous tassels being male and providing pollen and the ovaries on the cobs from which the kernels of corn will develop, being female. A single strand of the silk which protrudes from the end of the ear is attached to each ovary and provides a conduit by means of which the pollen falling on it fertilizes the ovary to which it is attached. Pollen is taken from the tassels to the silk by wind, and the individual plants cross-pollinize freely. Unlike squash or melons, in which cross-pollinization with a related species will not show up until the resulting seeds are planted the second year, cross-pollinization in corn may show up the first year. Field corn or pop corn, for example, that crosses with sweet corn growing nearby, will harm the sweetness, flavor and texture of the immediate crop when the ears are picked. Further evidence of the cross would be seen if the seed were planted the next year. Even a small amount of crossing between varieties of sweet corn planted close together will show up in inferior flavor.

For the vegetable garden, acceptable quality may be maintained by planting an early variety between two late varieties, or vice versa. This arrangement, however, will probably not be adequate to maintain a variety if one is saving seeds. To maintain purity in a strain, it is advised to segregate varieties that mature pollen at the same time by a distance of a quarter of a mile from each other. Maintaining a strain requires some ingenuity and manipulation, and probably cooperation from gardening neighbors as well.

For both sweet and dry corn varieties, seed collection is the same. Depending on the variety of the corn and on the weather, it requires about six weeks after pollinization for corn seed to mature. To save seed, sweet corn is grown as it normally is for eating, but instead of being picked while it is in the milk stage it is allowed to mature and harden on the stalk. Ears may be allowed to hang on the stalks until late autumn, or in the old-fashioned way mature stalks may be cut and stacked together in the traditional picturesque shocks. If the ears require additional drying as they usually do after being removed from the stalks, they may be husked and allowed to dry where there is good ventilation and protection from the weather. The old-fashioned method of pulling back the husks but leaving them attached to the cob, then braiding them together to be hung for further drying, is as effective as it is charming. When fully dry the kernels are not difficult to remove by hand. Some gardeners prefer not to keep for seed the kernels on the tip of the ear, from the point at which the ear begins to taper into a cone, feeling that these seeds do not have a good rate of germination and do not grow strong and productive plants. The same procedures for collecting and harvesting apply to both sweet and field corn, although the kernels of sweet corn will appear withered and smaller than field corn, as well as usually being much paler.

Carrots have perfect flowers and are cross-pollinized by insects. Plantings separated from each other by about two hundred feet will retain relative purity, but if a high degree of purity is required a separation of no less than one thousand feet is required. Carrots will also cross with Queen Anne's lace, a close relative, and this plant should not be allowed near where carrots are being grown for seed.

Carrots are biennial, producing seed their second year. Some individual carrots in a given planting will grow seed stalks their first year. This is an undesirable characteristic, and such plants should be pulled and discarded as soon as they appear. When digging the carrot crop in the fall, select and lay aside individuals with characteristics desirable in the variety. After storage through the winter, reset these roots in the early spring, planting them to the same depth they were when they were dug. Soon the carrot will put out a number of leaves, then grow a central stem to a height of three or four feet. A number of lateral stalks or axials will grow from the main stalk, with a flowering umbel on the tip of each. When they begin to form and dry in the late summer, the individual heads may be clipped and laid aside, protected from the rain, to complete drying. When we grew carrot seed on my family farm we cut the stalks with corn knives and left them in windrows, and when they were thoroughly dry we collected the seed with a very small combine used as a portable thrasher. This taught me that when dry the stalks shed their seeds easily and must be handled with great care. Carrot seeds have a strange pungent smell that I find curiously stimulating, I think because the smell powerfully and on a completely primitive level, evokes pleasant recollections of childhood.

Potatoes are almost always propagated by clones. Individual tubers are cut into chunks, each piece containing one or more eyes, as the buds on the tubers are called, and these sets, or "seeds," are planted to grow the new crop. This reproduction is vegetative rather than sexual: there is no exchange of germ plasm, and therefore the progeny will be genetically identical to the "parents." Because of this there is no advantage to selecting as seeds superior individual tubers. There are stories of frugal gardeners or farmers who year by year ate their big potatoes and kept the small ones for seed. According to these stories, eventually the only potatoes their plants produced were the small and insignificant potatoes called in Ireland "scideens." Likely there is an element of truth to the stories, but not a genetic truth. If potatoes are planted year after year in the same place with no fertilizer, eventually the only crop will be scideens. If these small potatoes were planted in good soil, no doubt they would produce normal crops.

Potatoes do produce a seed pod, which very much resembles a small green tomato. I once read that every summer somewhere in the country a news story appears featuring a potato with tomatoes growing on the tops of the vines. The "tomatoes," of course, are the potato's seed pods. A nice story, I thought, but I could not conceive of anyone being actually that naïve about potatoes. A short time ago, however, in midsummer, I saw featured on a TV newscast a proud if somewhat mystified gardener showing off his potato vines that were growing tomatoes. In full color, the gardener was pointing to a nice but quite ordinary potato fruit. Occasionally seed may be found in these pods, but more often they will contain nothing but unfertilized ovaries. If seed is found and planted, most likely it will not be exactly like its parent, and it may be in fact quite different. A correspondent from Sweden commented in the 1982 *Seed Savers' Yearbook* that because of a famine in Sweden in the nineteenth century people saved the seed from the potato pods for replanting and ate their seed potatoes. This introduced a greatly varied genetic stock, and for this reason, according to that commentator, Sweden today has an extremely large number of potato varieties.

In 1981 a new variety of potato was introduced, one grown from true seed rather than clones, a variety that has been bred to throw true to type from seed. The advantage is that it is cheaper and easier to keep and to transport seed rather than whole potatoes or sets cut from them. To mature full-sized potatoes from this seed in northern areas, however, they must be prestarted inside, then transplanted to their permanent spot after the weather has settled.

Most gardeners who plant seed potatoes from their own crop simply sort out at planting time small potatoes that have been among those stored for winter from the previous year's crop. The only selection necessary is for disease or rot. Any potatoes that show scab or fungus infection should be discarded to prevent transferring the disorder to the new crop.

A gardener who is growing and saving his own seed should be concerned with storing the seed in such a way that it will indeed grow the following years. Once seed has been matured, harvested, dried and cleaned, how long the seed will remain viable—capable of sprouting—will depend to a large extent on how it is stored. We all know as a matter of practical experience that just "keeping" seed, as distinct from deliberately storing it, works perfectly well on a short-term basis. In any given year most of us find several partially filled commercial seed packets with the opened top carelessly double-folded back, and we plant the contents with results that are completely satisfactory. Similarly, the dried cantaloupe, lettuce, broccoli or whatever seeds we save can simply be placed in an envelope in a drawer or box, and they will almost certainly grow the following year. With a little care, however, the seed can be made to keep and remain viable for a much greater length of time. This becomes particularly important if one is trying to maintain a large number of different kinds of seeds and lacks space or time to grow them all every year. For species for which it is difficult to keep varieties pure without cross-pollinizing, such as cucurbits, brassicas, corn, etc., if seed can be sustained several years one can plant a single variety each year and keep the strains perfectly pure.

It is most important that for long life seeds be kept cool and dry. Therefore, seeds must be *thoroughly* dry before being placed in storage. Seed savers are fond of saying that when one thinks his seeds are quite dry enough, let them dry a week or so longer. If it seems that takes too long, remember: what's time to a seed?

After seeds are thoroughly dried, to maintain low-moisture content pack them in sealed jars together with a substance which absorbs moisture from the air. The two agents most commonly used are powdered milk and silica gel. The material is put into a cloth bag, or folded into a cloth wrapping, and placed in the storage jar. If more than one kind of seed is to be stored together, segregate seeds according to their planting time so that they may all be removed and the remainder returned at the same time, to avoid excessive opening and closing of the jar. If powdered milk is used as the desiccating agent, use a volume about equal to the quantity of seeds to be stored. A note in the 1982 *Seed Savers' Exchange Yearbook* explains that silica gel is much more effective in absorbing moisture than powdered milk. A spoonful will do the same job as a cupful of powdered milk. Silica gel can be obtained in hobby shops, camera shops, drugstores, florists', etc., and is not particularly expensive.

Small quantities of dried seed may be enclosed in an envelope, with all pertinent data written on the outside—name, characteristics, breeding background, date, etc.—and placed in the container with the disiccant. If this is done, the life of any given seed should be several times longer than the normal estimated on the tables of viability for seeds simply stored in envelopes.

The seeds for beans and peas need to be handled differently. It is advised that bean seed be first put into a tight container to protect it from moisture and then frozen for twenty-four hours. This freezing will kill weavil eggs which may be under their skins. Peas and beans should not be stored in airtight containers, but kept in such storage as will permit them to "breathe." Cloth or burlap makes good storage for them. These sacks should be kept where it is cool and dry, and should be stacked or hung in such a way as to permit air to circulate around them.

From time to time one reads accounts of seed taken from ancient tombs—Egyptian, Sumerian, Aztec, etc.—which after centuries and centuries has been planted and grown. Remembering such accounts, one might conclude that if seeds are so tenacious, why make such a mystery out of saving them for a mere five or ten years? It is true that some seeds are very long-lived, and that by scientific testing some viability has remained in certain varieties beyond a century. However, I have read several times that it is highly doubtful if there has been a single case of a truly authenticated germination of a so-called "tomb seed." While I have known seed exchange members to offer seeds that they claimed were descended from tomb seed, without authentication I am inclined to doubt such claims.

My experience with carelessly handled seed has not been encouraging. In 1980, I found in the bottom of an old box a packet of poppy seed which I had collected, my notation on the packet said, in 1958. I planted perhaps five hundred seeds from this trove, and not one germinated. Perhaps if they had been packed as recommended above, at least a few of them might still have been viable. So by my ignorance in the past that strain is lost to me. A great pity, since it was a lovely huge blue poppy with single petals, the cultivation of which has, I believe, since become unlawful.

4. THE HEIRLOOM ORCHARD

Heirloom apples are the fruits that get the most attention from guerrilla gardeners. An appreciable number of people, who are interested in preserving old strains and who offer them for sharing or exchanging, present lists of varieties of scion wood that are impressive indeed. Identifying varieties and establishing their ages and origins is often a problem for amateurs. A number of state universities preserve and propagate collections of old fruits with records as complete as any available, and they are helpful in making identifications. In addition, some university collections will, to a limited extent, provide scion to amateurs. A number of references helpful in identification of heirloom fruits are listed in Appendix III.

One of the interests of those who fancy old varieties of fruits is to discover and identify specimens still found growing in old orchards, farmyards, abandoned fields, etc. Many collectors travel great distances seeking out and examining specimens in the same way buffs seek out historical sites and archaeological remains. Besides enjoying the excitement of the hunt and the discovery, and unlike comparable hobbiests, these collectors are actually able to take a tiny piece of twig with them and repropagate the object of their quest on rootstock in their own plantings.

While many old varieties have been preserved, many others have been lost, and these are the varieties collectors hope to locate, identify and preserve. In addition, many varieties of old fruits survive in local cultivars which have developed adaptations for local conditions. Among those interested in collecting specimens there is considerable correspondence, exchange of information and exchange of scion. Most of these "plant explorers" also belong to one or more of the societies, or subscribe to and are listed in the publications, of the guerrilla gardeners. A number of such organizations are listed in the selected bibliography of Appendix III.

There are many reasons an individual gardener might become interested in heirloom fruits. An obvious one is dissatisfaction with fruit that commercial considerations, and not quality considerations, have engineered for the marketplace. Probably this is the major reason for people to "drop out" of the commercial supply chain and turn to the home garden. And it is certainly a major reason why additional gardeners are considering turning from conventional suppliers of commercial seeds and nursery stock. It is a logical second step from determining to grow one's own fruit and produce for the sake of freshness, etc., to deciding to grow the varieties one wants even at the expense of some effort in obtaining them, rather than being dictated to by the commercial supply of plants in the marketplace.

Many people remember varieties of fruits and vegetables from golden periods of the past. Large numbers of seed and scion requests in the Seed Savers' Exchange and similar publications give a name or a description of something remembered from times past that the member would like to retrieve or sometimes just identify. Or one hears one's elders speak with relish of a variety or varieties remembered from their youth. In addition, some such varieties have traditionally had only local distribution, and one wishes to experiment with extending their ranges.

One need not have any reason specifically in mind, or any variety specifically in mind, and may be motivated by nothing but curiosity and the spirit of adventure. Whether they admit it or not, this has to be a contributing motive to all who cultivate unconventional species and varieties. There is something of a sense of gazing through a window into the past as one grows an apple hardly seen since the time of one's grandfather. Or of William Shakespeare.

These latter interests have unquestionably been my major motivations for growing heirloom fruits. It is basically a romantic sense of the past that makes me want to grow varieties of apples (and roses, for that matter) mentioned by Shakespeare and Chaucer. But I also have a sufficiently practical curiosity to discover whether some of the almost fabled varieties of apples are indeed as good as they are fabled to be.

There is little point in offering a "suggested heirloom orchard," or recommending varieties that should be given a try. The following description of varieties has no logical rationale other than that all are old, interesting and possible to obtain, though not necessarily through conventional means. The list of varieties discussed is short, especially considering the magnitude of options; anything even approaching an ambitious, not to mention exhaustive, list would not stop short of one hundred varieties, and if it were truly ambitious, or were to attempt the impossible idea of completeness, it would include many hundreds of varieties.

The apple called the Arkansas black, mentioned earlier, is an heirloom variety which probably still has limited commercial, or locally commercial, cultivation. It has in the past been used as a pollinizer for other commercial varieties of apples and may or may not have been picked for retail. The apple is large, half a pound commonly, almost perfectly round, and of an especially beautiful red-black color. The flesh is pale yellow, very crisp and quite tart. It is a winter apple, keeping easily through February. In fact, the flesh is so firm that it is difficult to eat out of hand until it has mellowed two or three months in storage. It is not especially difficult to find nursery stock.

Much more a candidate for consideration on the basis of antiquity is the Ribston pippin. It has been a standard dessert apple in England for many centuries. It is not especially pretty, being described as bright orange with a touch of red, and with brownish-red netting. The flesh is hard and extremely sugary, with a fine rich flavor and aroma. It ripens from late September to December and keeps fairly well. Henry Leuthardt describes it as the most exotic apple worth growing in America. My tree produced its first fruit, a single apple, in the fall of 1982. I left it for a few days in our kitchen for the privilege and honor of being in the presence of this fabled apple, and my wife, not knowing what it was or why it was there, decided it was too ugly to keep around and threw it away.

The northern spy is an apple with origins at least in the middle of the nineteenth century, and as a name it is familiar to many people. It is a large, red winter apple with good keeping qualities. It has considerable fame as a pie apple. At least one author has speculated that the "spy" in the name may originally have been "pie." My impression is that it is not commonly grown other than by fanciers, but a number of commercial nurseries stock it. It has the disadvantage of being rather slow to come into production.

The smokehouse apple is an old variety; according to Southmeadow Fruit Gardens, it originated in 1837 in Pennsylvania, and it still seems to be most commonly grown there. It is a sweet apple, red and striped, and has an excellent reputation as an ingredient in cider. It also has a reputation for being a dependable producer.

The Cornish gilliflower is an English apple dating from the eighteenth century. It was once described by a great English pomologist as the best apple known, with a firm yellow flesh, rich flavor, extraordinarily sweet juice, and a sweet fragrance described as clovelike, hence its name. It is not a pretty apple: dull green, with dull reddish-brown striping or webbing. It is also oddly asymmetrical in shape.

The Spokane Beauty must be a commoner apple than I thought. Since it was first brought to my attention, I have seen it offered by several nurseries. It is reputed to be one of the largest apples known, with weights of two and a half pounds said not to be uncommon. I have found little, if any, comment on its eating or culinary properties, which may be significant. It is to me, anyway; I have two of them, one of which was sent to me as an unsolicited bonus with another order, and neither has yet produced.

The Spitsenberg is an apple one often hears the older generations speak of with fondness. It seems to have been at one time an important commercial apple, which has lost out to considerations of glamour in merchandising. It is characterized as having a rich, spicy flavor and a distinct aroma. The color is deep yellow with a spreading red blush. Many people rank it among the best for eating. While it is not commonly seen, some commercial nurseries offer it.

The court pendu plat is among the oldest apples known. It was possibly introduced into England during the Roman occupation. Under nearly a hundred different names, it is cultivated all over Europe. It is a flat, squat apple that grows tight against the branch, more like a peach than an apple. Its color has been described as being like Italian marble, a basically fawn-colored skin blushed with yellow or orange and a touch of rose. The flavor is described as deliciously musky. In England the variety is commonly called the "wise apple," because it blooms so late that it misses the late spring frosts. It also ripens its fruit very late. This is a very rare apple. The only listing I know for it is Southmeadow Fruit Gardens'.

The Kerry pippin, as its name indicates, is an apple of Irish origins, being first noted at the beginning of the nineteenth century in Kilkenny. It is a small yellow apple, sometimes with striping, and has a crisp hard flesh and a fine rich flavor. Nitschke characterizes it as the best late-August apple. It is uncommon. The only listing for it I have seen is in Southmeadow Fruit Gardens.

Lord's seedling is that rarity of fruits of any kind, an amusing apple. It was sent by a man named Lord to the New York Experiment Station in 1892. Although quickly determined to have no commercial potential, it has so charmed generations at that station that no one ever had the heart to discard it. The late Professor Howe, co-author of the series "Fruits of New York," described it as "one of the most aromatic, deliciously flavored apples I know. As a commercial variety it is absolutely worthless." The apple is not attractive, yellow with russet mottling, and ripening is in late August. It is described as a dependable and heavy bearer. Such a mutt of an apple as to have absolutely no commercial value has to twist a few heartstrings. To my knowledge, it is offered commercially only by Southmeadow Fruit Gardens, although I have seen scion offered in non-commercial listings of the exchanges.

The Cox's orange, or Cox's orange pippin, is given about as high a praise as an apple can get from the experts. It is a quite old English variety, although curiously it seems to be more difficult to grow in England than in America. It has a tender yellow flesh, said to leave an aftertaste like ice cream on the tongue. In *The Apples of England* it is flatly called the greatest apple of our age. It is also an attractive apple, red-orange to bright red. It ripens in late September and is supposed to have moderately good keeping qualities.

An American apple that has virtually disappeared from America but is almost a standard in England is the mother. This is a red dessert apple with a yellow sweet flesh, a distinctive flavor that has been described to be like pear drops, and an aroma compared to wintergreen.

Another American apple, which Nitschke characterizes as "the classic American apple," is the Newtown pippin, also known as the yellow Newtown. I think care should be taken to determine variety by description rather than by just the name, because a so-called yellow Newtown common where I grew up was nothing like the one described by Nitschke. The Newtown pippin dates from the middle of the eighteenth century. It is green, sometimes with a hint of red, and is a long keeper. It is said to come into its own in March and April after winter storage.

A fruit once immensely popular but today seldom seen other than as a flowering dwarf, is the quince. In the past, however, the quince was sometimes esteemed as a preeminent luxury fruit. Its center of culture was the Mediterranean area, especially southern France, where numerous varieties were grown. So highly regarded were the quinces of France that in the fourteenth century foreign ambassadors to the King of

France were regularly greeted at the border with a basket of the fruit. Considering the limited appeal and culinary use of quince today, one may suppose the fruit to have then had qualities since lost. Of the few recipes one can find today even in old recipe books, probably the one most likely to be familiar is quince jelly. I remember it vaguely from childhood as a very pretty greenish-orange jelly with a slightly astringent flavor that I did not particularly care for. The fruit for this jelly came from a tree in my grandmother's orchard, which was considered to be something of a curiosity and oddity in the area at the time. The ripe fruit was round and about the size of an apple, with a greenish cast and a distinctly fuzzy skin. Uncooked it had a flesh that lacked crispness and had a somewhat stringy texture and an alumlike flavor. Since I cannot conceive of such a fruit being so prized that French kings singled it out as a horticultural jewel of their country, I can only suppose that such superior varieties have been lost or have disappeared from common cultivation. It could be a worthy project to try to recover the lost honor of the quinces of the French kings. There seems to be a slight interest in this fruit within the gardening underground, and two or three varieties are offered occasionally by commercial nurseries. Unless one of them is by chance the fruit of my grandmother's orchard, I would not be familiar with any of them. If the old varieties have been lost, it is a good example of the ultimate result of the diminishing gene pool of a species.

Like the quince, and as the preponderance of their names suggest, the majority of the old varieties of pear are French in origin. Pears were introduced into England by the Romans, and there is an Anglo-Saxon word for them adapted from the Latin. They do not seem to have become popularized until after the Norman Conquest, when no doubt new and probably better cultivars were introduced from France. The fruit itself is ancient and, like such other ancient fruits as the apple and the pomegranate, occurs as a motif in mythology. There are persistent traditions which associate the pear with lechery. It was under a pear tree that Priapus attempted to ravish a sleeping nymph, who was saved when she was awakened by the laughing bray of an ass. The pear is also one of the many fruits conjectured to be the forbidden fruit Adam and Eve ate in Paradise. Chaucer describes a lurid and baroque sexual encounter taking place among the branches and green fruit of a pear tree. Chaucer also describes an especially ravishing young country wench as being as lovely as a "pere-joinette" tree. There is even speculation about a lost or forgotten significance that is not quite proper in the lover's gift of a partridge in a pear tree in the somewhat ambiguous traditional Christmas carol.

Also somewhat ambiguous is the piece of popular wisdom which states:

> He who plants pears
> Plants fruit for his heirs.

On the one hand, it may be cheerfully saying that since a pear tree is noted for its long life, if you plant a pear tree you will be sharing something with your descendants. You can almost hear some young nipper a few generations down the line with hair the same color as yours offering a golden fruit to a lovely young girl with the enticement "Great-Grandpa planted this tree." But pears also are sometimes exasperatingly tardy coming into really bountiful production. Somewhat less cheerfully, then, this conventional wisdom may be interpreted as cautioning that when you plant a pear tree you're planting fruit for your descendants because it certainly won't bear fruit during your own lifetime. Lugubriously, the imagined explanation of the future might well be, "It looks as if finally Great-Grandpa's pear tree set a few blossoms."

Whichever meaning actually brought this little couplet into being, neither applies if one plants dwarf or semidwarf trees. They should most likely begin bearing fruit about three years after they are set out. And most likely they will not live nearly as long as standard pear trees. Your great-grandchildren will bloody well have to plant their own.

Pears are somewhat fussy about pollinization. Many are

self-fruitful, but some have fairly precise requirements for pollinization. Even self-fruitful varieties, however, will pollinize much better and be much more prolific if there is more than one tree growing, to allow cross-pollinization.

A serious difficulty of growing pears is an infection called fireblight, and some areas are particularly prone to this infection. Fireblight is a somewhat enigmatic viral disease. Its name comes from its major visual symptom, a browning or blackening of leaves at the growing tips of branches, which gives the appearance that they have been touched by flames. The symptoms move toward the main trunk, with the twigs and branches quickly withering and dying. Standard treatment is to prune away the affected limbs behind the blight, cutting into healthy wood hopefully to prevent the infection from spreading to the main trunk of the tree. The importance of sanitation is emphasized in this treatment: take care to remove and burn everything pruned away and to disinfect the tools used in pruning. It is also important that infected material be cut away and destroyed as soon as it appears. Use any commercial sealant to treat the wound and prevent moisture loss, and to help prevent reinfection.

The best defense against fireblight is prevention. In areas where fireblight is known to be a problem, there may be serious and distinct limitations to the varieties one might plant. There are some pears that are highly susceptible to fireblight, some that are resistant and some that are virtually immune. A county extension agent, a state agricultural college or a local nursery man should be consulted about potential fireblight problems in a given area before selecting varieties. Most nursery catalogues in describing a given pear will comment on its resistance to fireblight or lack of it. When dealing with heirloom varieties, the information may be harder to come by and may be available only through the experience of other growers of heirloom varieties.

As more becomes known about the infection, probably better methods of prevention and control will become available. It is known, however, that too-rapid growth may trigger an outbreak. For this reason anything that stimulates quick growth should be avoided. Fertilization should be very modest and is best restricted to low-yield organic treatment, such as mulching with grass clippings. Also, any severe or radical pruning after the first two or three years is apt to induce rapid growth of suckers on the inside of the tree, and this rapid growth seems to invite attacks of fireblight. It is best to restrain growth to a modest and not luxurious annual increase.

There seems to be less active interest in antique pears among heirloom gardeners than in apples. Perhaps a reason is that the varieties of pears grown even in most commercial plantings commonly are old varieties, although new cultivars are introduced every year. The preponderance of French names, if nothing else, is a clue to the heritage of pears. There are, however, no doubt many possibilities in exploring this species for rare and ancient varieties as well as for little-grown gourmet varieties. I have heard, for example, of Japanese pears fabled for their excellence. Increasingly, Oriental pears are appearing in listings, and this is certainly an avenue of exploration well worth fuller investigating. And what may be found in old orchards, particularly considering the longevity of the trees, turns up interesting varieties. I have planted, for example, a pear that Leuthardt offers under the name of Atlantic queen, which is supposed to have descended from scion of an ancient East Coast tree of French, but otherwise undetermined, origins. My tree has not yet come into production, but the nursery described it as exceptionally hardy as well as excellent in the quality of its fruit. I have planted another variety offered by the name of Bennett from the Cloud Mountain Farm, descended from a West Coast pear grown in the Puget Sound area since the 1880s. It is described as similar to the Bartlett, but is vaguely characterized as better and, perhaps more important, resistant to fireblight.

The Bartlett must be in America at least the commonest and best-known pear and the one most often purchased either fresh or canned. It is also a home-orchard favorite. While it is self-fruitful, like all pears it produces better crops if there is at least one other pear tree nearby for a pollinizer. Unfortunately it is quite prone to fireblight, and this susceptibility should be taken into account before selecting it for one's orchard.

Pears are one of the few fruits that are better if picked green than if allowed to hang and tree-ripen. If they are left to ripen on the tree, flavor and texture suffer. Some varieties, including the Bartlett, develop an unpleasant graininess, almost as if sand were imbedded in the flesh. Some varieties develop soft, watery centers, and some tree-ripen with a flat, insipid or foxy flavor. Although picked while the color is still grass green, Bartletts happily are one of the easiest to ripen to their proper flavor. Pick them when they are still dark green with at most a stippling of lighter color, in late August or early September, depending on the area. If stored in the dark with a cool to cold temperature they will keep several weeks. Bring them to room temperature with normal light, and within a few days the fruit should ripen uniformly to a golden yellow. Once ripened they are very perishable and need to be used quickly.

Pears are notorious for their nonconforming shapes, which are ongoing exasperations to those who would pack and merchandise the fruit. There is something almost playful about the ways they don't conform. The conventional round shape of many fruits—apples, peaches, plums—easily accommodates to varieties of convenient and efficient packaging which features the ideal of commercial packing, conformity. Pears don't conform. "Pear shape" is almost synonymous with deviation from any acceptable geometric form. The availability of particular kinds of pears in retail outlets is remarkably proportionate to just how conforming a particular variety is and how resistant it is to symmetrical packing.

One of the most notoriously unconforming varieties of pears is the bosc. Some people uncharitably say that this pear is ugly. It resembles a caricature of Cyrano de Bergerac's nose, and its color is a dull or mottled dark brown, with a somewhat leathery skin. Pear fanciers designate its shape *calebasse.* Obviously if it did not have a truly exceptional flavor it would have long ago become extinct. It is a quite old variety, and although it is sometimes found in markets it is increasingly becoming a rarity, and it is worthwhile to have it growing in a home orchard. The fruit should be picked when the stem easily breaks loose from the fruit spur when the pear is lifted. The bosc keeps well in cool dark storage. As well as being good eaten fresh, it is a favorite cooking pear.

Similarly, the name "winter nelis" suggests that it is a pear that will keep well. It is another old-fashioned variety, becoming less commonly found in ordinary retail markets. The fruit should be picked late, in September or October, depending on the particular region. In suitable storage it will keep a good part of the winter. Traditionally it is cooked, although it is also good fresh. A somewhat newer pear developed from the winter nelis is the Michelmas nelis. It is said to be somewhat larger than its parent and to ripen better.

Of the so-called dessert pears, several receive competing mention for being the *crème de la crème* of all varieties. One often mentioned is the seckel. In its way, this is almost as unlikely-looking a fruit as the bosc. It has a teardrop shape, with a speckled brown or russet skin; the fruit is hardly larger than a walnut. Seckels are a dainty little fruit that do commendably when served with a quiet elegance. They are hardly the fruit of choice to a hungry hand reaching for a handful-sized portion of food. Their size reflects unfavorably on their commercial prospects all the way from picking and processing to merchandising. They are seen occasionally in markets, especially as a holiday novelty item, but in general they are commercially uncommon.

Of the dessert pears probably the one most often seen in the markets is the Anjou. It is a fruit with a chastely rounded pear shape. Feelings about this variety are somewhat mixed. Its commonness probably results more from qualities desirable commercially than from its inherent virtues. Like the red delicious apple, it offers no surprises: attractively shaped and colored and easily recognizable. I do not feel it is an especially good choice for the home orchardist. While it keeps well, I have always had trouble getting it to ripen properly. It is picked when the skin is of a pale-green color. It is ripe when the stem end of the fruit yields to firm pressure of the thumb. The skin is thin and easily damaged even before the fruit ripens, and when picked it must be handled with delicacy to prevent bruising or tearing. Small abrasions resulting from rough handling lead to brown spotting or decay when the fruit is stored and ripened. Properly ripened, the flesh has a soft creamy consistency and a light sweet flavor, but the flavor lacks distinction or character.

Another pear designated by some as the best pear of all is the comice, also called the doyenne du comice. The fruit is large, many commonly weighing over a pound, with a short neck and a broadly flared bottom; bell-shaped, perhaps, rather than pear-shaped. It matures in late October and does not keep especially well. It should be picked when the skin is still a pale green and when the stem breaks easily from the spur. This large pear, properly ripened, I serve sliced in half, with the small core removed, to be eaten with a spoon like a cantaloupe. Smaller specimens are excellent served as wedges, with the small seed cavity removed. Unfortunately the comice is susceptible to fireblight. My single tree, after several years of fighting the infection, probably will not survive much longer.

Another dessert pear labeled as top of the heap by some is the Flemish beauty. It is described as rich and musky in flavor, with a balance between sweet and sour. The tree is also very hardy. The fruit ripens in late September and early October.

The French, whose judgment in pears should not be taken lightly, favor the passe crassane as the best of all dessert pears. It is described as quite sweet, with an indescribable elegance of flavor. The fruit should be picked as late as possible, and allowed to ripen indoors very slowly. It is said to be at its best about March. It has a reputation of being a little demanding as a garden variety, requiring a richer soil than is usually served up to pears.

The Sheldon is another highly praised dessert pear. The ripe fruit is juicy and buttery, with sweet and fine flavor. It comes highly recommended for garden planting. I have a tree which has been an outstanding producer. I have not been troubled by fireblight in it, although I have fought a losing battle with the infection in other pears growing nearby. The fruit is picked at the very end of October.

The rousselet de Rheims is an ancient French pear, once the favorite of Louis XIV, for whom it was grown in his garden at Versailles. Reflecting its spicy, musky flavor, it is also called the spice pear or the musk pear. It has a most enticing perfume as well.

There are a number of Oriental varieties of pear which are little known in America, but there seems to be a growing interest in them. Southmeadow Fruit Gardens describes one that it calls simply "Chinese pear," but it is not exactly identified as to variety. It is characterized as having flesh crisp as an apple's yet paradoxically almost fluid in its juiciness. For one interested in a rather complete departure from familiar pears, this or other varieties of the Oriental pear would bear looking into.

Clearly, a home gardener with any sense of adventure at all would find among the varieties of pears not usually found commercially, some that would be irresistible. There is no more reason a home gardener should grow nothing but Bartlett pears or Anjous than he should grow nothing but red or golden delicious apples. Tasting the old, rare and unusual varieties of pears is almost like discovering a complete new species of fruit.

Although the apricot is an ancient fruit, originating probably in Turkey or Iran, the possibilities of growing old or gourmet varieties remain relatively limited. The word "apricot" is cognate with the word "precocious," a relationship arising from the fact that this fruit ripens very early in the season, considerably earlier than its look-alike cousin the peach. In his play *The Duchess of Malfi,* John Webster hinges a crucial portion of the plot on the fact that gardeners used to hasten the earliness further by ripening green apricots in the heat of composting manure.

Apricots are another fruit which the fruit explorers are investigating to locate older and superior strains. To a great extent our pioneering ancestors propagated apricots by the simple expedient of planting seeds. Unlike some fruits, apples and pears for example, fruit from seedling apricots is seldom worthless. Most frequently the fruit of seedling apricots is small and the seeds are large, but often though it is small the fruit is sweet and flavorful. Seedling apricots have another advantage over the seedlings of most other fruits: they also come into bearing precociously. A seedling apricot germinated by the edge of my father's compost heap in the mid-1950s. After making growth the first year that was more like a woody annual than a tree, the second year it blossomed and bore fruit. The third year it produced a sizable crop. A seedling apple, on the other hand, probably wouldn't have produced for ten years; one authority estimated for apple seedlings an average of seventeen years from germination to production.

Oddly, the pits of many seedling apricots are quite edible, with a sweet, almondlike flavor. There is a persistent belief that their pits are poisonous. Since childhood, when I ate seeds every year from my grandmother's seedling tree, I have tasted and tested seeds from other apricot trees without finding one that was inedibly bitter. If ingested in quantity, they might even be poisonous as well as inedible; I don't know. However, as a boy I ate those seeds from my grandmother's apricots, and I may even have felt guilty not to have died when my elders scolded me for eating something everyone else knew to be poisonous. Today I feel somewhat vindicated to find that an apricot being marketed by the name of sweetheart, developed by Stark Brothers Nursery, is especially advertised for the edible kernel.

The Moorpark is an old variety of apricot that is still commonly offered by commercial nurseries. Its name comes from an English estate named Moorpark, where it originated in the middle of the eighteenth century. It is a very high quality fruit—sweet, luscious, large and attractive. When it is ripe its dark orange skin is freckled with brownish dots that one might first mistake for insect damage. The Moorpark is at its best tree-ripened; picked underripe for the commercial market it tends to ripen into a pale orange, and its flavor is decidedly insipid.

The Blenheim, also named for the estate in England where it was first grown, dates from the early nineteenth century. The skin and the flesh are orange, and as with the Moorpark there is a freckling of red dots. It is a common commercial variety.

The Nancy is an old variety, the ancestor of the Moorpark. It is sometimes called the peach apricot or the pêche de Nancy because of its size. It is pinkish-orange in color and is described as rich in flavor.

The moniqui is a relatively unknown apricot which has been highly praised in professional publications in France. It is described as a large fruit with a white skin spread with a reddish blush. It seems to have been widely grown in Spain. Unless some of the fruit explorers have it, it is apparently not yet grown in America. However, Southmeadow Fruit Gardens indicated they would soon be offering it.

Southmeadow also indicated they intended to push research on registering and propagating old varieties of apricots. No doubt the fruit finders are doing the same. It's to be hoped that previously unknown or obscure varieties may soon be available to those interested and curious enough to look for them.

Next to apricots, less seems to have been done with investigating and preserving old varieties of peaches than with other fruits. One reason for what might seem to be neglect is that peaches are notoriously short-lived. On the other hand, around old farms truly ancient pears and apples are common; whether identification is authentic is doubtful, but in England the tree reputed to have borne the apple that was so inspirational to Isaac Newton is still shown. Even the brittle apricots sometimes endures an astonishingly long time, surviving with an appealingly frail and fragile look. One I remember feeling sorry for as a child is still producing small sweet fruits and, I think, is privately feeling just a wee bit sorry for me. But peaches seem locked to an unyielding time scheme: young, mature and gone.

Several of the popular varieties of peaches are, at least by some standards of heirloom gardening, antiques. The elberta is still commonly thought of as the aristocrat of peaches by many gardeners, although fanciers would not necessarily concur. The elberta has the advantage of large size, pleasant pale-pink color, an excellent flavor and good canning quality. If it has a disadvantage, probably it is that the fruit ripens late. In fact, its maturity is so standard that nursery men sometimes use the time of its ripening as an index against which to judge the ripening time and relative earliness or lateness of other varieties: i.e., plus or minus so many days before the elbertas ripen.

There are, however, other and older varieties commended by the heirloom gardeners, varieties which are largely uncommercial. Perhaps the oldest American variety grown, dating from the beginning of the nineteenth century, is called the George IV—which is an odd name for an American fruit. The George IV is characterized by Nitschke as the most delicious peach he has ever tasted. It is smallish and has a greenish-white skin. Nitschke describes its flavor as a perfect blend of sweetness and acidity, juicy and aromatic. It originated in the early nineteenth century, and during the time it has been around some of the most prestigious experts on fruits have given it their highest rating. This would seem to be a good candidate for someone interested in exploring possibilities unavailable in fruits for commerce.

Another old American variety is the late Crawford. Since it ripens at the end of September, its timing is welcome. The fruit is quite large and, unlike some heirloom peach varieties, is a vivid yellow. It is described also as having an excellent flavor and an unusual texture, firm with a melting buttery quality when eaten.

Probably the most ancient American peach is the Oldmixon free, believed to have developed from a clingstone peach planted in the early eighteenth century by the English historian Sir John Oldmixon. It was developed in later generations into a freestone. The fruit ripens late, in mid-September, and it has a pale blushed skin and white flesh. The flavor is described as rich and sweet, and many peach fanciers have given it extravagant praise. The tree itself is also distinguished by its hardiness.

The slappey is an antique peach now pretty much in the hands of the fruit collectors and connoisseurs, although it was grown commercially at least as late as the 1940s, and perhaps later. The peach is shaped somewhat more flat than round, like an exceptionally plump small discus. The skin is unusually thin, and it has no, or almost no, fuzz. From what I remember about growing the fruit commercially, it was troublesome to harvest because it bruised easily while being picked, and pieces of skin frequently slipped when it was being handled by the packers in the warehouse. Its shape also made it difficult to achieve the symmetrical uniform pack in the peach boxes that distributors find so desirable. Although it has commercial disadvantages, it has qualities that appeal to the fancier. Because of its lack of fuzz and its thin skin, it is especially desirable for munching. It has an excellent and exceptional flavor, sometimes described as pinelike or herbal. It also has a distinguished reputation as a favorite of home canners. The fruit ripens somewhat earlier than, or at the same time as, the elberta, in late August.

The grosse mignonne is a very old European peach, dating back at least to the seventeenth century. It is known in England as the Grimswood's Royal George. It is a fairly large peach, described as having a greenish-white skin, sometimes yellowish, sometimes mottled with red. The flesh is white and juicy, with a rich flavor. It is a late peach, ripening in early September.

The stump-the-world is a peach with an appearance as uninspiring as its name. Its color is pale green occasionally dressed up with a bit of a reddish blush. But it was long ago observed that its good qualities more than compensated for its looks. It is juicy and tender-fleshed, with a sparkling rich flavor and an excellent aroma. It is similar in qualities to the Oldmixon, but extremely late, ripening in late September. It is an American variety over a hundred years old. It's now extremely rare, being cultivated only in orchards of connoisseurs.

Although many different plums are commonly found in produce markets in season, the varieties found will almost exclusively be modern cultivars bred particularly for their commercial values, particularly shipping qualities and attractiveness. For the same reason that the delicious apple dominates home plantings, these commercial plum varieties are those most commonly offered by nurseries. Most modern gardeners are not even aware that other varieties exist. Because many people like to have a couple of plum trees for the pleasure of having plums tree-ripened for eating out of hand, considerable interest should develop in heirloom plums as their value becomes recognized. The epicurean varieties are distinctive and, because of their fragility and perishability, have never been a practical item of mass commerce. They require tree ripening, and when properly ripened they are beyond anything but the most deferential and delicate handling. Plums we remember being praised in works of literature were most likely grown in the cottage and manor gardens close to where they were eaten or, as Robert Browning mentions, in the gardens of the cloister ("As he gathers his green gages, Ope a sieve and slip it in.")

The greengage is among the most celebrated of the antique plums. In France it was named the Reine Claude, honoring the wife of King Francis I, and it will occasionally still be found referred to by that name. Its English name honors Sir Thomas Gage, who first brought the plum to England. The fruit is small and green, and it ripens in September. It is said to be so sweet that when fully ripe the flesh will semicrystallize, often rupturing the skin. Neither its appearance nor its handling qualities would make it possible for the greengage to survive as a commercial variety.

There are several other varieties of gages which are very old. The Reine Claude diaphone, also called the transparent gage, is larger than its cousin the greengage, and it too ripens in September. Its flavor is ranked very high in quality. The Reine Claude violette, also called the purple gage, is another ancient French variety. The skin is dark purple or purplish black, with juicy yellow flesh. It has been called the best-flavored of the purple plums. It is best if left hanging on the tree until the fruit begins to shrivel like a dried prune. It ripens in mid-September.

There are quite a large number of other excellent varieties of antique plums available if one takes the trouble to seek them out. The most interesting of them seem to be the European varieties. Since plums grafted on such dwarfing rootstock as *Prunus besseyi* (as supplied, for example, by the Southmeadow collection) may be maintained as hardly more than bushes, most gardeners should be able to afford the space to experiment with several varieties.

Cherries have become a much more possible fruit for the home gardener since the availability of dwarf and semidwarf stock. Standard cherry trees may get very large, requiring the space of two or three standard peach trees. If only standard trees were available most gardeners would scarcely be able to afford space for one.

The advantages of growing one's own cherries obviously include first of all that of having tree-ripened fruits. Like peaches and apricots and the better varieties of plums, cherries are fragile and do not ship well. Therefore to have first-quality cherries one must grow one's own or procure them from someone who does.

Cherries are not especially difficult to grow except in the respect that the sweet varieties are sensitive to cold. In northern areas with extremely cold winters the trees will winter-kill. They are also vulnerable to spring frosts during blossoming time. Occasionally hard freezes that occur after the trees begin to break dormancy, but before the buds open, will kill the buds and they will not blossom at all. In commercial orchards such expedients as smudge pots, gas jets, and fans are used to cut the edge off spring frosts and protect the buds and blossoms. For the most part the home orchardist in areas of marginal climate is limited to planting his trees in protected locations in which the lay of the land prevents collection of cold air into frost pockets.

Sour cherries in general are not nearly as tender or susceptible to freezing or frost as sweet cherries. Some of the sour varieties, for example the North Star, can survive and flourish in about the harshest winter experienced in the lower forty-eight states, and are easily adaptable to a Zone 4 area. One year the Zone 5 area in which I garden experienced a record cold spell during which the temperature reached forty-five degrees below zero in late December. All of my deciduous trees and shrubs were heavily damaged, and only one of three apple trees I then had survived. My North Star cherry tree was unharmed, and it produced its usual crop of pretty fruit the following summer. This variety, however, is extremely sour, and even when dead ripe the fruit requires considerable sweetening.

Most produce markets offer only a limited variety of sweet cherries, and these varieties are pretty much mirrored in the varieties offered by nurseries. However, older and more unusual varieties may be found through the exchanges as well as through certain commercial sources. These heritage cherries present considerable variety in color as well as in flavor.

Among the most interesting and certainly the oldest of cherries are the European varieties. The early purple gean is a very early and large heart-shaped cherry. The name comes from the French *guigne,* a cherry that is soft and juicy. This term is opposed to bigarreau, the name of a cherry with flesh that is firm and crisp. The early purple gean originated in the early nineteenth or late eighteenth century. Its flavor has been described as a kind of sweet and sour.

An example of the crisp-fleshed or bigarreau type is the Merton bigarreau. It is not an especially old variety, having been developed in England in the middle of this century. The skin is dark red, the flesh firm and crimson-colored. In England the fruit is highly recommended for home gardens. Although a rarity in this country, it is reported to do well where grown. Southmeadow calls it consistently the richest-flavored of the red cherries they grow.

What may be the oldest variety of cherry under cultivation is the yellow Spanish, which is believed to be a fruit described by Pliny, the Roman naturalist who died while investigating the eruption of Mount Vesuvius. In England it has been called the graffion. The color is yellow and crimson, and the flavor is highly rated. The tree itself is reputed to be exceptionally vigorous, which it certainly must be to have survived so long.

The early Rivers is described as the best of the English sweet cherries, dating from the 1870s. It has a dark-red or reddish-black color, but it is uneven in shape—which would be enough to kill its chances in the American commercial markets. However, it has a singular quality which should make it tempting to commercial growers: it is reputed not to crack in wet weather, as such commercial varieties as bings do with exasperating certainty if rain occurs while they are ripening.

The English Morello is likely to be familiar to some gardeners, at least as a name. It has been around since at least the sixteenth century, and was mentioned by Shakespeare. The fruit is tart, but when it is fully ripe its flavor is excellent. The cherries are dark red and extremely late in ripening, being harvested in late August. When cooked they are described as having an especially attractive color and aromatic flavor.

The spate braun is a German cherry from the Rhineland. It is a sweet cherry with a crunchy-crisp flesh and excellent flavor. The fruit is large, with a dark-red skin. It is also exceptionally resistant to cracking. Nitschke judges it the best of all cherries of its size and texture.

The Kirtland's Mary is an American variety from the first half of the nineteenth century. One authority called it the finest-flavored of all cherries. The color of the flesh is distinctive, pink or pinkish rose as distinct from red.

The range of possibilities for old and noncommercial varieties of cherries is much larger than this list, although not nearly as staggering as the possible choices of apple varieties. It should be readily seen that besides the adventure and novelty of growing heirloom or "connoisseur" varieties, together with the advantages of obtaining tree-ripened fruit, there really are, with all varieties of orchard fruits, enormous ranges of flavor possibilities to which almost no one outside the fraternity of the guerrilla gardeners is exposed.

One last genre of fruit, bearing no resemblance to anything grown in almost any home orchard, deserves a mention for the sake of its historic and linguistic peculiarity if nothing else. This fruit, the medlar, is virtually unknown in this country today, and it is probably not much better known in Europe. Yet literature abounds with references to it, persistently sly or even lewd references. Modern literature students are absolutely stumped by such allusions as Shakespeare's respecting things "which maids call medlars when they laugh alone." Or Chaucer's use of the more than earthy country name for the fruit, "open-ers," calling it a fruit that is worthless until it has rotted in the straw. The fruit is mentioned or its name punned on surprisingly often in literature, and never without a snicker or a raising of the eyebrow.

The fruit is indeed an oddity. It resembles a large rose hip having a prominent and enlarged calyx, slightly opening as it ripens to reveal parts of the first row of seeds inside. It is picked in the fall after having been frosted, at which time it is still green and inedible. It is placed in cool storage (perhaps, as Chaucer advised, laid in straw) until the fruit softens, at which time it turns a pale brown. Writers sometimes note or allude to a distinctive musky odor associated with the ripe fruit. Most uncharitably, one medieval writer characterized it as not being worth a turd until it was ripe, and then it tasted like shit.

The tree itself is reputed to be very handsome, and pictures I have seen of it tend to confirm this. A distinctive feature of the tree is that it branches oddly, with limbs forming sharp right angles to the main stem. The medlar is really more of a shrub than a tree, growing only to a height of about ten or fifteen feet. While it is unlikely at this late date that the medlar will make a comeback as a fruit, we can at least hope that one will be planted prominently near the English building of every university and college.

5. GAMING WITH THE ICE SAINTS

Most Temperate Zone fruit trees experience a period of dormancy which must be undergone or they will not fruit properly. The dormancy required varies widely and is usually expressed in terms of the number of days or hours when a particular species must experience a certain temperature or below. Only after such a period of dormancy will fruiting buds begin to develop as spring warming occurs. This requirement means that such cold-hardy trees as apples, which have fruiting buds that mature after winter weather is over, can survive severe winter temperatures and fruit normally, providing that freezing does not occur after the buds begin to develop. Good crops are thus determined less by the lowest winter temperatures than by consistency of mild weather at blooming time. But these requirements also mean that an apple, for example, that does well in a Zone 6 area which is frost-free after early April will likely also do well in a Zone 5 area that is not frost-free until early May. But the same apple may also be vulnerable to frost in both zones.

In Devon, England, which has enjoyed centuries of fame for its cider, legend has it that Saint Dunstan once attempted to establish a beer business in the ancient and mystic city of Glastonbury. He found this to be an unprofitable undertaking, however, because the natives preferred to drink their own excellent domestic apple cider. To undercut the competition, Dunstan is reputed to have made a pact with the Devil to provide frost in the apple orchards at the most critical time of blossoming. The Devil agreed to provide Dunstan with killing frosts on May 12, 13 and 14. Since in the calendar of the Church saints these days were dedicated respectively to Saints Pancratius, Servatius and Bonifactius, these men have come to be called the "Ice Saints." Whatever the connection or cause and effect, unseasonably cold nights occur with exasperating consistency around the time of the Ice Saints, and orchardists continually hope to pull their orchards through this critical time in the development of their apple blossoms. The ancient court pendu plot apple has been called the "wise apple," because it has the cunning not to blossom until after the days of the Ice Saints.

Because their dormant buds are more fully developed and exposed than those of apples and pears, the stone fruits are more vulnerable to frost after they open, even after they are pollinized and the fertilized ovary begins to develop into a fruit.

Consequently, a person when planning an orchard needs to take his climate into account more than with any other part of his garden. Too-mild climate means there will not be a sufficient period of dormancy for many kinds and varieties of fruit trees to blossom and bear. Even for such cold-demanding fruits as apples and pears, however, varieties which will produce in most southern zones have been developed. As well as general climate, in planning to plant fruit a gardener should also take into account his own microclimate—how the land he has available is situated, how it lies, and which parts of his garden may be marginally more secure from frost than others. While this latter point may be an easy one to state, of course most people have to make use of whatever land they have. Far more people have available at most a few hundred square feet than have several varied acres among which to make their choices. However, what limited choice one has should be used.

Since cold air sinks and slips beneath the warmer air, if possible avoid planting fruit trees in so-called frost pockets: low areas which do not have "drainage" to allow cool air to pour away rather than collect in cold basins. Most suitable is a sloping field with a southern or southeastern exposure, which both drains cool morning air and allows warming at the earliest possible hour of the morning. However, in clearly marginal areas there is an advantage to a northerly or westerly slope because such a situation remains cooler than a southern or western exposure. This will hold back the development of the buds a little longer so that hopefully they will not be far enough along to be vulnerable until after the days of the Ice Saints have been safely passed. It is better to plant fruit in a situation that will cause the crop to be consistently late than in a situation in which the fruit will be early (on the lucky chance it doesn't freeze out altogether).

Most fruit trees require a well-drained soil to do well. If the roots are constantly in strata of saturated soil, trees will often die after reaching a certain age and size. If the drainage problem is simply that of a heavy soil that is by its texture and consistency too water-retentive, when planting dig the hole especially wide and deep. Fill the bottom (depending on the severity of the condition) with a foot or so of small gravel, then fill in the planting hole around the roots with porous material rather than with the soil that was removed in digging the hole. If the problem results from the land lying low in such a way that it collects and retains water, probably the only solution is to arrange a system of draining ditches or tiles. There are flexible porous pipes that may be installed, in complex patterns if need be, to collect and remove excess water from even rather troublesome areas.

The spacing between trees depends primarily on the expected size of the mature trees. Even when trees are planted on rootstocks that produce standard-sized trees, authorities vary widely in their recommended spacing. Commonly, commercial plantings are spaced wider than home garden plantings. For standard apples, a spacing of about twenty-five by twenty-five feet would be acceptable, although much wider spacing may be recommended. If dwarf trees are planted, the degree of dwarfing will vary with the dwarfing rootstock, and the instructions frequently provided when nursery stock is purchased often make specific recommendations for the particular rootstock onto which the trees are to be grafted. Here also rather wide differences may be seen in nursery recommendations even when the same rootstock is used. With the common dwarfing rootstock designated Malling Merton 106 (MM106) (see the rootstock table, page 119) a spacing of twelve by twelve feet should be adequate; that is what I have come to accept, and I find it generally satisfactory. Unless special pruning or training is planned, trees should not be spaced much closer, although some people who grow larger numbers of different varieties do space much closer because they plan to prune and train the trees almost as if they were shrubs.

Fruit Trees: General Information*

	Height	Years to Bearing
Apples, semidwarf	12–15 feet	2
Apples, standard	20–25 feet	3–10
Apricots, dwarf	8–10 feet	2
Apricots, standard	15 feet	3
Cherries, sour, dwarf	8–10 feet	2
Cherries, sour, standard	20 feet	3
Cherries, sweet, semidwarf	15–20 feet	3
Peaches, dwarf	8–10 feet	2
Pears, dwarf	12–15 feet	2
Pears, standard	30 feet	3–8
Plums, dwarf	8–10 feet	2
Plums, standard	20 feet	3

* After *Burpee's Nursery Guide.*

When newly planted, the young trees will seem especially far apart, but after only a couple of years the trees will be seen quietly occupying their territory, and the gardener may begin to wonder if he hasn't after all planted them closer together than he should have. If crowded, trees will not achieve a proper shape and consequently productivity. Particularly as they grow larger and begin competing for light and space, the trees will attempt to make up in height what they lack laterally. This tends to produce tall scraggy trees which are difficult to harvest and spare in production. I have made the mistake of planting too close together, and to my embarrassment some of the results still remain accusingly around my premises.

When selecting fruit varieties for planting, it is necessary, often crucial, to select varieties that will complement each other for best pollinization. Some varieties are self-fertile, but some varieties are self-sterile. All will make better production if more than one variety is planted. With some varieties the problem is that when a blossom opens it will release pollen from its stamens (the male, pollen-producing parts) before the pistil (the female part, connected to the ovary) has matured to a readiness to be fertilized. With self-infertile varieties, therefore, not only is it necessary to provide a second parent tree, but the second must also be a tree which matures pollen at the time the pistil of the first tree is receptive. Nurseries indicate varieties that are especially particular, and what varieties they will pollinize with. Although not always noted by nurseries, some varieties have a tendency to fruit every other year, regardless of the weather and the season. If one does not know of this habit, it can be most frustrating to find a season that has seemed perfect produce a crummy crop, and a seemingly impossible season produce a fine crop.

In the recent past the almost universal practice was to transplant all trees, shrubs, etc., in the early spring, just before the plants began to break dormancy. Increasingly, however, nursery stock is being shipped and transplanted in the late autumn, after the leaves have fallen but before the deep freezing sets in. The rationale for autumn transplanting is that the tree will experience two cool seasons before the first hot weather. Also by spring the soil will have completely settled around the roots, and the healing process of the roots in their new growing medium will have been completed; thus, transplant shock may be reduced or eliminated altogether. Some gardeners maintain that it is possible to get the effect of nearly an additional year's growth by autumn planting. Many nurseries now regularly supply stock for autumn planting and even publish special autumn catalogues of nursery stock. Autumn planting should be done before the ground freezes solidly. It is a good idea to prepare the planting holes in advance, as early as mid-September, and wait until virtually the last minute to transplant.

When nursery stock is ordered, whether for autumn or spring planting, most nurseries supply both one- and two-year-old plants, and often older trees as well, even some which are too large to be shipped through the mail or by parcel service. Usually the one-year-old plants are somewhat cheaper. Probably they will be under two feet tall and will have a correspondingly smaller root system. Some gardeners maintain that besides being a bargain, since they are smaller the one-year-olds transplant with less shock. Because of their size some say they should not be pruned at transplanting, so that they will make a taller growth their first season. According to this notion, shaping by pruning should begin the second year, having at that time a stronger and taller main stem to work from. I have used both one- and two-year-old transplants, and although I can't say I feel there is a sharp difference, I still feel more comfortable with two-year-old stock for fruit trees.

Shipped from the nursery, the trees will arrive bare-rooted. This means that all soil will have been removed and the roots usually packed in a light absorbent material such as sawdust, wood shavings, etc., that has been moistened. They should also be sealed in moisture-retentive bags. Some nurseries do send out their bare-root stock bare indeed—I have received them without so much as a protective bag—but I have had almost no success with stock which arrived in that condition.

Bare-root trees are very sensitive to exposure to air and to heat. Once unpacked, they should not be exposed to the air even for a brief time before transplanting. If a number of trees are packed together it is best to have all holes dug and prepared before removing the protective covering. The roots may be further protected by submerging them completely in water until they are actually placed in the ground.

In preparation for planting, in normal soil the hole should

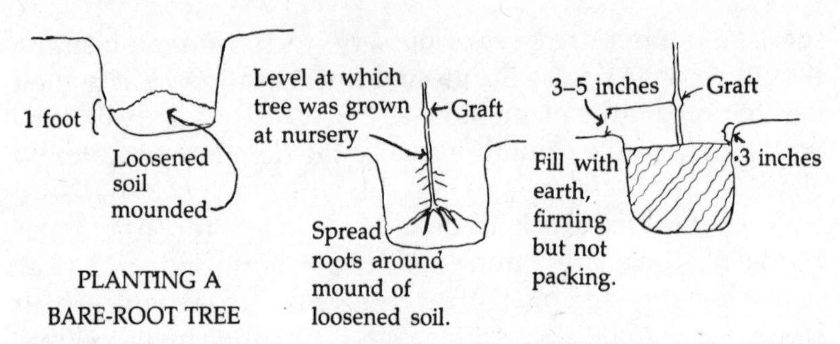

PLANTING A
BARE-ROOT TREE

be dug at least a foot deeper than the actual bottom level of the roots, and at least a foot wider than the distance the roots will actually spread, which will almost always make the hole at least three feet in diameter, probably more. If the soil is extremely heavy, dig the hole deeper, and add a foot of light gravel. If the soil is extremely sandy or otherwise excessively permeable, add about a foot of sphagnum moss, and mix it into the bottom soil.

The ideal hole will provide loosened soil at the bottom and the sides, providing a loose medium into which new roots can easily penetrate as growth begins. A tree planted in a cramped hole may show good growth the first year or two and subsequently never be especially thrifty. I have seen trees which were doing badly after having been in the ground for several years and which, when they were dug up, were found to have their roots mostly knotted about the area of the original too-constrained hole.

Whether or not gravel or sphagnum has been added, there should be at least a foot of loosened soil under the roots, which will allow easy penetration for the deep-seeking roots. Make a conical mound of soil at the bottom the diameter of the hole, and about eight inches high. This will provide a pedestal upon which to rest the base of the rootstock, and a plane about which to train the lower roots. The actual depth of the hole, and height of this earth cone, will finally depend upon the size and shape of the individual tree.

Nothing should be added to the hole before the tree is set—no manure, commercial fertilizer, etc. It is possible that fertilizer touching the roots may kill them. It is possible, too, that the presence of fertilizer or other enrichment may stimulate too much top growth and make the top outgrow the ability of the roots to supply moisture, making the young tree vulnerable to midsummer heat. It could also result in growth that is weak or excessively leggy, from which it might take the tree a year or more to recover.

When the tree is unpacked and brought to the hole, use the tree itself to make final adjustments for depth. The point of the tree from which measurements are made is the bud or graft union. This union is readily discernible as a scar, a point where the stem changes thickness, where there is a change in the texture of the bark and usually a slight irregularity in the straightness of the stem. The tree should be set into the hole to such a depth that this graft union is about three inches above the surface that the soil will finally make around the tree. If the union is too high above the soil, there is danger that there will be a structural weakness at that point in the mature tree trunk. If the union is set below the surface of the soil, the stock above the graft union will root, and if the tree is a dwarf, the dwarfing value of the rootstock will be lost, resulting in a standard-sized tree.

When the depth of the hole has been properly adjusted to fit the individual tree, settle the bottom of the tree on the top of the soil cone, spreading the bottom roots evenly around it. The hole should be wide enough that the end of the roots do not reach closer than six inches to the outside. If there are any broken roots, they should be pruned between the break and the stock. Sift soil around the roots, firming it but not tamping it. If there are roots growing from the stock above those spread across the bottom cone of soil—as there probably will be, as soon as the soil reaches them—spread them as they might have been when the tree was originally growing.

Continue filling the hole with earth, firming but not tamping, until the soil is to within about two inches from the top. From a large bucket or slow-running hose add water until the soil is saturated and water is standing on top. The soil will settle at least a couple of inches, and more soil should be added until there remains a depression of about three inches from the top of the surrounding soil. This will serve as a reservoir for watering until the tree becomes established. By the end of the first year it should have been filled in until only a shallow depression around the base of the tree remains.

If there is a secret to successful transplanting of bare-root (and other) trees it is to keep them well watered until they are fully established. The day after transplanting I always provide each tree with enough water, added slowly, that water remains standing in the depression. I give that much water every two or three days until leaves begin to form, and I continue to water the same amount about twice a week until new growth is forming and the shoots are a couple of inches long. Then I continue to water the same quantity once a week for the entire summer, until it is time to begin hardening them off for winter. I live in an area with little or no summer rain; in a wetter climate less water will be required.

After watering, the next most important thing for the survival and future health of the tree that can be done at the time of planting is to prune them. Many people find that this is psychologically a difficult thing to do. One does not want to sacrifice the least bit of a tree or cause it any more pain or trauma that it has already undergone. I keep finding running through my head the lines of the old song, "You sure be a stranger, and in a strange land," as I set out trees that were taken from the soil twenty-five hundred miles away such a short time ago. Just to put it to bed and to pamper it is the normal and human impulse. However, not to prune is doing the tree no favor at all.

Trees are pruned at planting time for two reasons: to reduce the amount of foliage the establishing roots must supply with moisture, and to induce the tree to begin a well-formed scaffolding, or structure, of branches. Those bare roots will have to secure themselves into the soil, grow hair roots and extend their range in order to draw all the moisture and nutrients

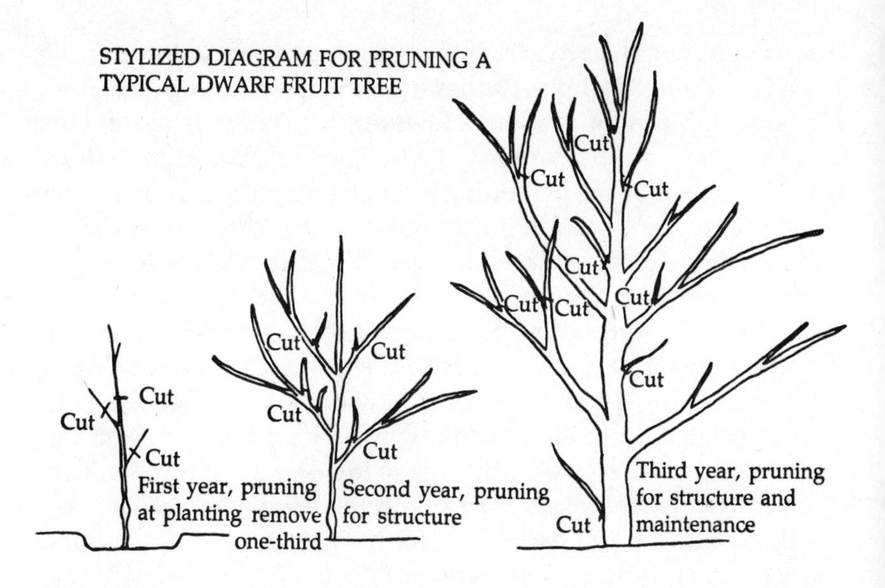

needed to become established and to strike a balance with the portions above ground. Too much top with too many leaves may cause the new tree to transpire more moisture than an unestablished root system can supply. Top branching may become weak, poorly formed or actually stunted; as the summer progresses and temperatures become truly hot and dry, the tree may die. Or it may be possible that the tree may end the summer so underestablished that it will not be able to survive its first winter.

The standard practice for pruning newly transplanted trees is to remove the top third. A tree which consists of an unbranched single stem or whip should be cut about a fourth of an inch above a bud. A shoot that should grow from that bud will become the new lead or main stem for the tree. What one wants to happen is for the tree to put out several shoots from which selections can be made the following spring to design a well-formed, sturdy and productive tree. If the tree does not consist of a single whip, but has several branches from its central stem, leave a total of three branches, each cut back a third of the way from its tip. The excess branches should be removed, cut cleanly at the main stem and left with no protruding stub. Nothing further should be necessary the first year in the way of pruning. However, if a tree fails to put out three good limbs, a limb may be induced to grow more or less where one chooses by cutting a half-moon notch through the outer bark to the cambium, a fourth of an inch above a leaf or bud. This should induce the tree to cause a limb to begin growing from that bud.

Two main styles of pruning shears, each with different form and different use, are available to the gardener. One is called a "bypass shear," with a design like that of standard scissors, in which two cutting edges slide tightly by each other. Because this design allows one cutting blade to rest against the trunk or lead from which the shoot or branch is being removed, the resulting cut will be nearly flush to the main stalk. If the shears are sufficiently sharp, no tearing of the bark will result; if they are not sharp, the bark of the main trunk might well be damaged. The second kind of shears is called an "anvil pruner." In this design, only one of the blades has a cutting edge, and it bears down against a narrow flat piece of metal, the anvil. Because this pruner provides its own solid backing against which the cut is made, it makes an exceptionally clean and neat cut, and unless the pruner is in bad condition it should never tear the bark. However, the width of the anvil and the frame to which it is attached holds the cutting blade away from the trunk or stem, and rarely can a cut be made closer than an eighth of an inch to the stem. Therefore anvil pruners always leave a stub. They are, however, usually quicker and neater than bypass shears, but they do have the limitation of not being able to make cuts that are flush. They are ideal for any kind of pruning where there is no concern about leaving a stub, but bypass shears or a saw are needed when a flush cut is required.

After transplanting, new trees should not be fertilized at all the first year. It may be useful but is not necessary to spread a mulch of some sort around the base of the tree—bark, lawn clippings, etc.—primarily to hold in moisture. The mulch should be pulled back several inches away from the trunk of the tree to prevent its being used as a gallery from which rodents may nibble at the bark of the tree.

For most varieties of trees a single light application of fertilizer may be applied the second year. A standard 10-10-10 fertilizer, put on at the rate of half a pound per half-inch trunk diameter, can be applied from spring to midsummer. It should be spread to about the edge of the drip line, or where rain would normally drip from the outer circle of leaves and onto the ground. The fertilizer should be worked into the upper level of the soil and watered. The trees should be fertilized at this same rate no more frequently than every two or three years. If fruit trees are overfertilized they will put on too much vegetative growth: the quick-growing long lush stems that disfigure the trees' conformation and rob growth and potential production from the main parts of the tree. Excessive vegetative growth tends to weaken a tree's structure and reduce fruit production.

Pears are very sensitive to fertilizing. If pears put out too much vegetative growth, susceptible varieties may contract fireblight, that serious and often terminal affliction of pears. A healthy pear tree is very long-lived and should be sustained by a slow, even growth pattern. A frequently recommended program of care suggests using nothing but an annual dressing of lawn clippings. In my personal experience I have planted and subsequently killed by overfertilizing two fine pear trees before I discovered my error. I now have two trees which are surviving well under a much more spartan lifestyle.

The second year after planting fruit trees, pruning should begin directing the tree to its desired shape. The most common basic plan for designing a tree is to produce a lead with three lateral or scaffolding branches. Select three limbs more or less on the same axis and as much as possible facing away from each other. Cut away any small limbs or sub-branches that grow inward or that cross each other. The ideal being worked toward is a tree that is sufficiently open and airy that fruit may size and ripen properly. Also it is of course desirable that the limbs be sufficiently uncrowded that they can grow strong and sturdy, not willowy and weak. The angle at which limbs anchor to the main stem is important in determining their strength. As one writer said, a carpenter's eye would determine the potentially strongest limb to be that with an acute angle, under forty-five degrees or so. However, a branch is not anchored to the trunk as a carpenter might anchor a brace. On a tree, a branch forming an acute angle is much more subject to tearing than one anchored at a wider angle. The ideal therefore is somewhere between a right angle and a forty-five-degree angle. The placement might be said to somewhat resemble the angle of a classically shaped or stylized menorah candelabrum.

A curious piece of natural history affirms a relationship between the veins in a leaf and the branches of the tree from which it comes. Examine a leaf and study the angles at which the veins of the leaf join the main vein of the leaf, then compare this structure to the structure of the tree from which it came. By and large there will be a very close correspondence to the angle at which the side veins attach to the main vein of the leaf and the angle at which the limbs attach to the main branch or trunk of the tree. I have never read or heard an authority suggest using the pattern of a leaf as a blueprint for pruning the tree, but the general correspondence can be easily confirmed simply by looking and comparing.

It is possible to worry a great deal more than is absolutely necessary about such things as the angles of the branches in one's trees. However, there are books and articles which treat in detail techniques of spreading branches by internal braces, external wiring patterns, etc., to produce ideal forms. Basically I have always just done the best I could with what I had. Even though I have made just about every mistake that is physically or horticulturally possible, using my somewhat casual and seat-of-the-pants methods I have never had a branch break because of structural unsoundness.

An alternate school of thought about pruning has recently received attention in various publications. Adherents of this practice advocate that a new tree should not receive any pruning at all its first year. Not cutting back, these practitioners say, stimulates the tree into earlier fruit production than if it had been pruned; the pruning sends a message to the tree's "control system" to replace damaged tissue rather than to prepare for fruit production. According to this theory and system, pruning—and that pruning light—should be reserved for the second year, with shaping being accomplished more through bending and bracing existing branches rather than through pruning away to selected branches. As attractive as the idea seems, my experience following it has not been encouraging. I planted two sweet cherries with no pruning and proceeded with only the lightest of pruning. They did indeed fruit quickly, having some blossoms the second year and a decent showing of fruit the third, a good two years ahead of my expectations. However, I have never had decent scaffolding branches to work with on either tree, and I'm resigned to the fact that the trees will never have proper shapes. Any gardener tempted to this system should remember that many nurseries state explicitly that all guarantees are canceled if a third of the tree is not removed at the time of planting.

6. DEALING "BETWEEN THE BARK AND TREE"

The great seventeenth-century poet Andrew Marvell wrote probably the most unsentimental and totally beautiful poetry about plants and gardens ever composed. Although any plant in Marvell's garden would be instantly and without qualification considered by a modern fancier to be an heirloom specimen, Marvell had already detected that things were going wrong in man's relationship with plants, that basically man was mucking things up with his meddling. He saw man taking upon himself the prerogatives of nature, virtually usurping the role of creator as he manipulated both the growing medium and the genetics of plants. Artificial arrangements in gardens Marvell did not mind; artificial soil and tinkering with the nature of individual plants, however, he abhorred and denounced. He held grafting to be the most sacrilegious of all of man's meddling. In "The Mower Against Gardens," after enumerating and chiding human transgressions against nature, he says:

> . . . yet these rarities might be allowed
> To man, that sov'reign thing and proud
> Had he not dealt between the bark and tree,
> Forbidden mixtures there to see.
> No plant now knew the stock from which it came.

Grafting, he says, is the ultimate offense against nature; it adulterates and bastardizes until nature no longer even knows itself.

How Marvell might wince and grimace if he could see the orchard of a modern collector of antique apple trees, where a dozen or more varieties might be grafted for the sake of preservation and perpetuation on a single tree, and the tree itself reduced to the role of nursemaid to a whole family of bastards. Practically speaking, no one can hope to obtain and propagate specimens of some of the rarer fruit trees without practicing rudimentary grafting. While a large number of heirloom trees can be obtained as one- or two-year-old transplants, especially by seeking out some of the less main-line nurseries, there are literally hundreds of varieties which may be located through the pipelines of the guerrilla gardeners. These are available only as scion, the piece of limb or twig which provides the living start from which one must bud or graft onto his own rootstock to produce a tree.

A graft consists basically of a piece of branch (scion) from a tree one wishes to grow, attached onto a root (stock) which will provide a suitable base on which it may grow. A graft is a clone: it takes living cells from a parent plant and perpetuates them. Since no sexual exchange occurs, in essence the parent plant itself is perpetuated. Although grafting may be done for a variety of reasons, the most common purpose is to propagate a desired variety.

The old American legend of Johnny Appleseed might lead one to suppose that all one need do to grow an apple tree is place an apple seed in the ground and wait. According to the story, that is exactly what Johnny did. After obtaining sacks of seeds, he wandered the countryside, periodically planting seeds as he walked along the roads. The happy conclusion was that after years of wandering he made America fruitful with apples.

Unfortunately propagating apples is not that simple. The vast majority of apples and other orchard and garden fruit do not "throw true" from seeds. That is, if you plant the seed of a winesap apple, the fruit borne by the resulting tree will most likely be inferior, even totally worthless, and may bear little resemblance to the parent tree.

In the countryside where I live, seedling apple trees are very common along the roads, grown probably from apple cores tossed from passing vehicles. Within a short radius from my home there are hundreds, perhaps thousands, of such seedling trees. I have examined fruit from enormous numbers of these trees and have found only two or three that were passingly edible. One, admittedly, was good enough that I made a graft of it onto a dwarf rootstock for my orchard. But, tellingly, there was only one good one out of all I examined. If you want a good apple tree, you must graft from a variety you know is good.

Although one cannot confidently propagate varieties of apples by planting seeds, there may nevertheless be just a shadow of practical truth in the Johnny Appleseed story. In early America it was indeed common to grow apples from seed, because the distance from the Old World made transporting living scion or trees almost impossibly difficult. No doubt seeds were simply planted from apples which had been imported from Europe. When in need, a crummy apple no doubt is better than no apple at all, and by the law of averages an occasional apple would be worth propagating in its own right. Today American apples are distinctly different from European ones; as a consequence of all this germ exchange, there are different genetic pools among American and European apples. This greater variety is thought to be a direct result of the genetic diversification produced through sexual rather than clonal reproduction.

However, it must have been recognized early in this country that if superior fruit was to be propagated it must be accomplished by methods other than seed planting. No doubt very soon many gardeners who were experienced in grafting began identifying and propagating desirable fruits wherever they might chance to find them. It is significant, though, and attests to the fact that the process was a slow one, that apples continued to be a popular import article from Europe through the end of the eighteenth century.

In 1847 a nurseryman from Iowa named Henderson Lewelling got the westering fever and joined a wagon train for the Oregon Territory. Instead of farm implements, his wagon was loaded with grafted apple trees. He caused considerable annoyance and aggravation to the wagon master because of his cargo's special requirements and the occasionally heavy drafts on the wagon train's water supply in the dry plains regions. But he managed to reach the Far West with about half of his trees alive. He established a nursery in the Willamette Valley in Oregon and provided the original source of Western apples for both personal orchards and commercial plantings. Descendants of his stock comprise a part of the heirloom fruit of the area today.

Not all fruit trees are as chancy to grow from seeds as apples. With apricots the likelihood of obtaining a worthwhile fruit from a seed planted by intention or accident are quite good. Many of the apricot trees found around old farmhouses are seedlings. The fruit tends, however, to have seeds disproportionately large for the flesh, and the fruit is seldom very large. Occasionally, though, one of these seedlings may be truly fine.

While many nut trees are regularly propagated from seeds, here, too, superior strains may be obtained with certainty only from grafts. A number of individual enthusiasts of the plant explorers scout growths of wild nuts in search of ones with superior qualities. Particularly sought are strains of pecans which will mature nuts farther north than their usual range. Also black walnuts and hickory nuts are sought for increased size of nuts, thinner shells, thicker meats, etc. Given the length of time to maturity and bearing age, so far it has been more practical to seek new varieties in nature than to proceed with orderly breeding programs to produce desired qualities, as do, for example, the tomato breeders. Once desirable features are found in a wild nut, it is hoped that many or most of the nuts produced from that parent will grow a tree having the same qualities. For absolute assurance that these qualities are transmitted, however, instead of the planting of the nuts, scion from the original tree should be grafted onto a new rootstock.

It becomes abundantly clear, therefore, that anyone who wishes to dabble in the cultivation of heirloom trees with the least degree of adventure or experimentation will have to be able to do rudimentary grafting.

It is the upward-growing stem of a young plant, not its roots, that determines the variety of the tree. In making a graft, one substitutes the growth tip of a variety whose fruit he wants for the growth tip of a plant whose root he wants. The genes that will make the new growth which will become the desired variety of tree reside in the buds of the scion. Within compatible species, it is relatively easy to force the stem of one plant to grow onto the stem of another. Making this happen is called grafting.

Fortunately, simple grafting can be made with good hopes of success by even the most inexperienced person. For the more advanced and those not content with the simple in anything, there are grafting techniques as well as purposes that are not particularly simple or easy. Anyone who becomes interested in advanced techniques should by all means consult R. J. Garner's *The Grafter's Handbook.* Many instructions for grafting, by the way, such as are found sometimes in government publications, are difficult to follow and, even when well illustrated, unclear as to procedures. The best advice when reading such works is that one should know what they mean beforehand and then not listen to what they say. Garner's *Handbook* is always clear, and what he doesn't tell, he says slyly, the reader will have more fun finding out for himself.

For most gardeners the first and perhaps only reason he will ever graft will be to propagate a variety that is not available as nursery stock from the regular sources of supply. Home-grafted trees are also a great deal less expensive than nursery stock, and for one who collects a lot of trees, that may become an important factor. For purposes of simple propagation it is necessary to plan far enough ahead to have the proper rootstock at hand when the time comes for grafting, particularly when one's scion is coming through the mail from a collector or other source at some distance.

Considered as a prospect for grafting, a tree is not always just a tree. It is perhaps self-evident that an apple, for example, may not be grafted onto a pear, or a cherry onto a peach. But cherries can be grafted onto certain kinds of plum, and apples can be grafted onto certain kinds of quince. More elaborate still, certain kinds of trees have been developed specifically to serve as rootstock for particular fruits and have no other purpose. In some instances where a preferred rootstock is incompatible with a desired scion, an intermediary is grafted between them, called an interstem, which is compatible with them both. It is like plating copper onto steel and chrome onto the copper because chrome won't plate on steel.

A grafter has three options in obtaining rootstock. One, he can plant seeds of the particular tree he intends to graft. Two, he may find seedlings of that particular species which have volunteered in or around his or someone else's orchard; sometimes a tree may have put out suckers far enough away from the parent tree to be used. Three, he may buy appropriate rootstock from a nursery. Some nurseries will supply seedlings of a particular variety to use for grafting. Some nurseries cater to home grafters' needs and supply a full line of rootstock for dwarfing, semidwarfing, growing standard trees, or

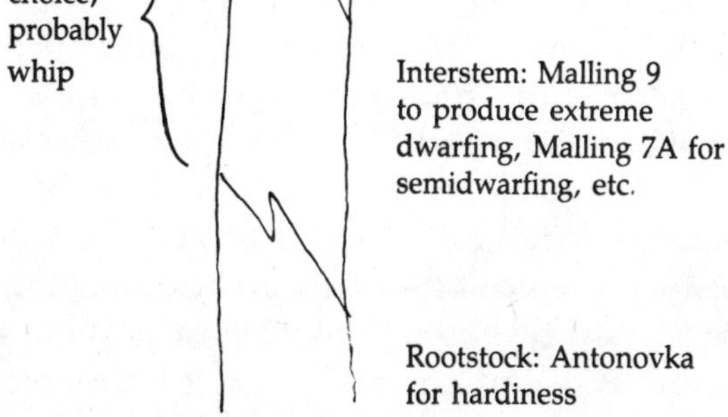

Scion *(apple variety of choice)*

EXAMPLE OF ONE USE OF AN INTERSTEM FOR GRAFTING AN APPLE

Graft of choice, probably whip

Interstem: Malling 9 to produce extreme dwarfing, Malling 7A for semidwarfing, etc.

Rootstock: Antonovka for hardiness

meeting particular requirements of climate, etc. Lee-Land Nursery offers a complete line of rootstock for all varieties of fruit. Particularly if bought in quantity, rootstock for grafting is not expensive, at least from the standpoint of the economics of the home gardener.

A part of a gardener's decision regarding what rootstock to use is determined by whether he wishes to grow standard or dwarf trees. A standard tree is one that grows to what we recognize as normal size. That's perhaps a question of relativity; if the mature tree is large enough to swing a hammock under, it's a standard. Otherwise, it's a dwarf. There is also a size in between the two that is called a semidwarf; you could swing a hammock between two of these, but not under one.

Some of the advantages of dwarf trees are obvious in the name. They use less space, and may be grown in a backyard where a standard tree would be too large or would hog all the space that might otherwise accommodate several small trees. Dwarf trees make it possible to grow several varieties in a small space, and because of their small size it may be possible to do other gardening between them. Small trees do not block views and do not block sunshine from other gardening. And, of course, being small makes them easier to tend: including spraying, pruning, thinning and harvesting. Another distinct advantage is that dwarfs come into bearing much sooner than standard trees; usually for most varieties a small crop can be expected the second or third year after being set out. In fact, the advantages of dwarfs over standard sizes are so numerous that many commercial plantings now consist entirely of dwarfs.

Grafts must be made upon a rootstock which is compatible with that species. While the rootstock must be compatible with the scion, it will not necessarily be the same variety of tree. Rootstock is bred and selected on the basis of its usefulness for dwarfing, for hardiness, for disease resistance, etc. Much of the rootstock used commercially and by amateurs as well has been developed by the renowned East Malling Research Station in England, and for this reason it is designated

Rootstock for Budding and Grafting Select Varieties of Fruits

VARIETY	ROOTSTOCK
Apple	Malling 9: This rootstock produces the greatest amount of dwarfing. However, it develops a somewhat weak root system.
	Malling 26: This rootstock produces medium dwarfing and develops a strong root stystem.
	Malling 7A: This rootstock produces semidwarfing.
	Malling Merton: This rootstock produces a sturdier tree than M7A.
	Malling Merton 106: This rootstock produces a standard-sized tree, but is valued for its extreme hardiness. For dwarfing it is grafted with an interstem of Malling 9, 26, etc.
Peach	Halford: This rootstock produces a standard-sized tree and is compatible with all commercial varieties.
	Siberian C: This is a dwarfing rootstock that is cold-hardy and not well adapted to warm climates.
	Tomentosa: This rootstock produces a tree about 40 percent the size of a standard. It also induces early bearing.
	Besseyi: This rootstock produces a very dwarfed tree, 30 percent the size of a standard. The trees are usually staked or trellised. It also induces early bearing.
Pear	Seedling pear: This produces a standard tree that is very hardy, and it is compatible with most cultivars.
	EM Quince A: This rootstock provides dwarfing and induces early bearing. It is not compatible with Bartlett, which requires an interstem.
Cherry	Mahaleb: This rootstock produces a standard tree. It is strongly rooted and hardy.
	Mazzard: This rootstock produces a standard tree. It is strongly rooted and hardy.
Plum/Apricot	Halford: This rootstock produces a standard-sized tree and is strong and hardy.
	Myrobalan: This rootstock produces a semidwarf tree and is quite hardy.

East Malling (or merely Malling) number 8, 9 or whatever. Amateurs who graft in some quantity often propagate their own rootstock by multiplying starts commercially obtained.

The bark of a tree consists of a dry, usually quite thin protective covering, with a thicker, moist and fleshy portion underneath attached to the woody stem. This second layer is called the cambium and is the outwardly growing part of the tree. Next year's growth ring is being formed by this year's cambium. When a graft is made, the cambium of the scion is brought into the closest possible contact with the cambium of the rootstock onto which it is being grafted. In a short time the two cambium layers will combine, and together they will produce a growth ring of wood, which will unite the stock and the stem into one tree. At the proper time for that species to break winter dormancy, the buds on the scion will begin to grow, developing into shoots. The best or most conveniently located of these shoots will be allowed to remain, becoming the main stem for the new tree. When it comes into bearing, the new tree which has developed will have the good qualities for which the parent tree of the scion was selected in the first place. It will be the original tree, of course—its clone.

Because it is vital that the cambium layers of the two pieces of wood unite as quickly as possible, grafting is commonly done in the early spring, just as trees are breaking dormancy, when the sap is beginning to run freely up through the cambium tissues, and healing and growth are most rapid. Certain kinds of grafting, especially for repair and strengthening, can be done on a more or less year-round basis. Scion for grafting is usually collected while it is in deep dormancy, and it is held in cool moist storage until the proper time to graft. It is important that the scion not be allowed to dry in the slightest, so it is often packed in moist sphagnum moss, frequently buried in the ground deep enough to protect it from freezing. Small quantities may be kept in a refrigerator.

Grafting may be done over a relatively long period, from when the rootstock is beginning to break dormancy, the point at which the sap has begun to swell the cambium, up until just before the parent plant begins to leaf. If one is ordering scion from some distance, difference in climate between the two places should be taken into account. Early spring in Georgia is deep winter where I live, for example. It is safest to order the scion in the late fall or early winter to be certain it will be on hand at the proper time for grafting.

There are an intimidating number of styles and kinds of grafts, but if all a given gardener wishes to propagate is a few rare or unusual varieties of trees he can obtain in no other way, he can easily get by using only one, the simplest kind of graft. This style is not foolproof or failproof, but it is easy to learn and it usually works.

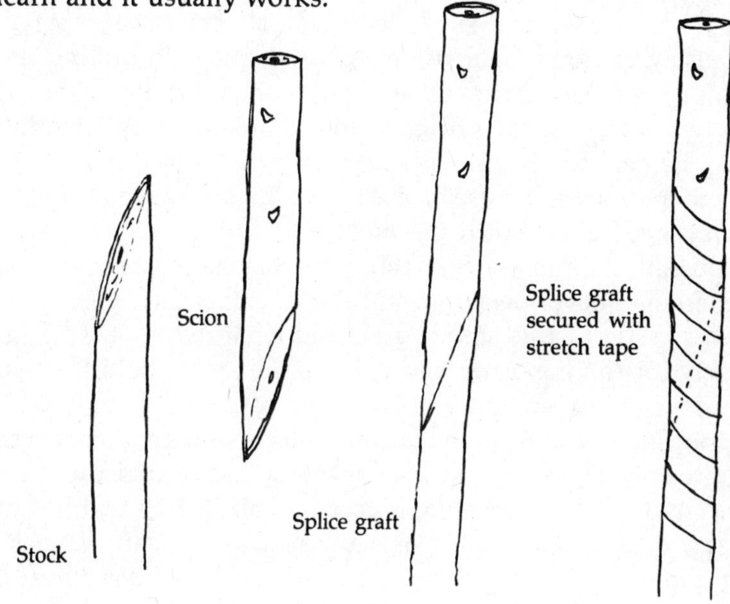

This simple graft is called the splice graft. For it to work requires that the stock and the scion be quite nearly of the same diameter. This requires measuring the two stems against each other and cutting them to match in size. Although there are a large number of different grafting tools and grafting aids available, one can get by with just a jackknife, provided the blade is extremely sharp. I have never used any other tool for grafting but a high-quality pocketknife that takes and keeps an extremely keen edge.

After the stock has been cut to length, make a long, clean diagonal slash that smoothly cuts away half of the stalk. The length of the cut should be about four to five times the diameter of the stock. Next, cut the scion with a long smooth slash identical to the cut on the stock with which it is to be joined. In making both cuts be careful not to fray or damage the cambium layer. An extremely sharp knife is essential in making a smooth clean cut with no ragged or torn edges.

Fit the scion to the stock. Do additional trimming as necessary to make the two sloping cuts match exactly. With experience and practice a single cut will almost always serve, requiring no additional trimming or shaping. In uniting the two pieces of the graft, you are mating the cambium layers of the two pieces of stem to grow into a union. Close attention must be paid to insure that when the two pieces are matched together these tissues are in exact and almost total contact with each other for the entire circumference of the union. It must be possible for sap to flow freely from the one into the other as if there had never been a rupture or interruption.

The secret of this simple graft and what makes it work so slickly for the inexperienced is the binding. To hold the two pieces together, serving the functions of securing the union, supporting the two pieces until they can support themselves by growing together, and of preventing loss of moisture, use stretchy plastic electrical tape, about a half inch in width. Cut off a workable length—six or eight inches—and, after attaching it to the stock an inch or so below the cut, wind it upward, pulling the tape tightly enough to make it stretch, and with a diagonal spiral wrap the full length of the union and about a half inch beyond. Be careful all the while to keep the two cuts and the two cambium layers in alignment so that total contact between them is retained the full length of the graft. When completing the wrap, be careful not to cover any bud of the scion with the top of the wrap.

Elastic electrical tape assures a tight fit to the union, and its moisture resistance also prevents dehydration. Its elasticity allows it to stretch as the diameter of the growing scion and stock increases. Before the availability of such tape, making a splice graft was rather chancy and finally more time-consuming than grafts which are sounder structurally but which require more experience and technique. Since both cuts of a splice graft are perfectly smooth, it is awkward to work with the two pieces of wood, keeping the cambium unions together and fastening at the same time. Obviously a helping hand could make things quicker and easier, but I have always done it alone and managed well enough, if clumsily. To make an amateur feel bad, Mr. Garner in his *Handbook* timed professionals working at East Malling and concluded that a person should complete a graft in less than a minute by himself, and in slightly more than half a minute if he had a helper.

A modification of the splice graft improves the ease of alignment and physical management of the two pieces. This design is called a whip-and-tongue graft, or simply a whip graft. Begin exactly as with a splice graft, selecting and match-

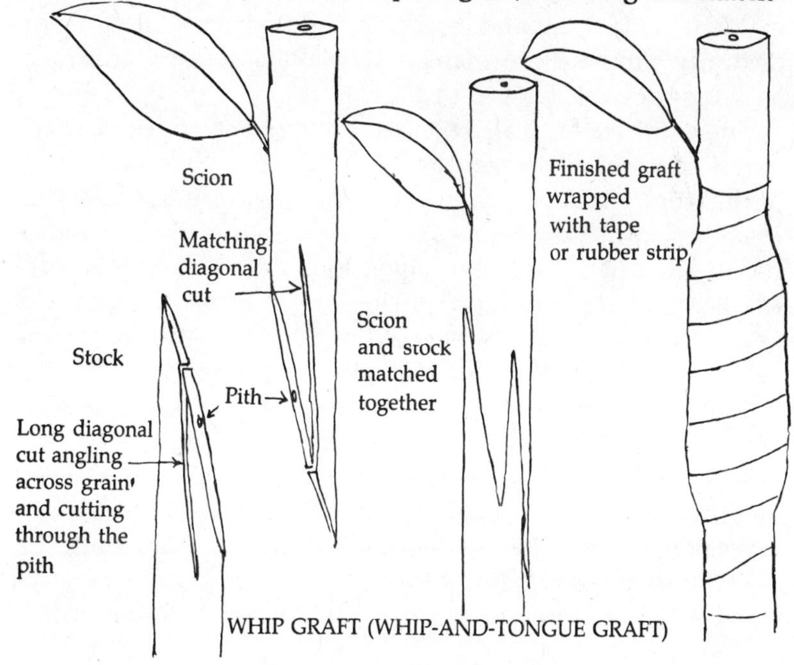

WHIP GRAFT (WHIP-AND-TONGUE GRAFT)

ing the diameter of the pieces to each other, and make a long sloping and matching cut through the stock and the scion. To make the "tongue," measure roughly halfway from the top of the sloping cut to the pith in the center of the twig and make a shallow cut angling across the grain, slightly below but paralleling the original diagonal cut. Stop when the new cut is halfway between the pith and the bottom of the original cut. Make a matching cut on the sloped surface of the scion. Note that it is not necessary to remove any wood. When I was a boy trying to learn this graft from a book, I thought it was necessary, and it almost drove me crazy. Fit the scion and the stock together, interlocking the two sets of tongues, working them together until there is a good matching of the cambium layers between the two pieces. Because there is an actual joining of wood in this graft, there is a much stronger union than there is with a splice graft. It may be strengthened still further by binding it with electrical tape, but most professionals would finish the graft by using a simple tie made with a length of rubber band and a dash of sealant to conserve moisture. It takes practice to make the tongues fit and interlock, and perhaps a bit of native dexterity. Anyone can manage to make a splice graft work, and it is probably best to begin with it.

The whip graft can also be used to unite a scion with a stock that is considerably larger than it in diameter. Remove the top of the stock with a flat, even cut, using sharp pruners or a saw. Make a slanting, diagonal cut down the bottom of the scion as before, as if it were to be joined to a stem of its same size. Make a matching cut up one side of the stock. To form the graft, cut a tongue on the side of the stock to match the diagonal cut on the scion. Fit them together, aligning the cambiums of both, and bind the union with tape or rubber bands. A sealant should be used over all surfaces, including the flat cut across the top of the stock. If the stock is fairly large, up to two inches in diameter, it is common to set two scions on opposite sides of the stem. The second year, the graft that has made the poorer growth is removed.

As the grafted trees begin to break dormancy in the spring,

one or more of the buds on the scion should develop into shoots. Remove any growths that develop on the rootstock as soon as they appear, in order to concentrate all of the energy in the new shoots on the scion. Occasionally with trees that make especially rapid growth, the shoots may grow so long in a short while that a heavy pressure will be placed on the point of union between the graft and the stock. If the new growth seems to be putting a strain on the graft, by all means support the shoot with a stake. Otherwise it will very likely break at the point of union, and the graft will be ruined.

A second way to propagate a desired variety onto a selected rootstock is by budding. Instead of uniting onto the rootstock a complete section of branch which has two or three buds, a single bud is removed from the scion (now called a "bud stick") and is attached inside the bark of the stock. There are many techniques and systems for budding, although not as many as there are for grafting. Budding also enlarges the season during which one may propagate trees. A large part of grafting is done before the trees break dormancy. Budding is done after the trees break dormancy, and some kinds of budding are possible from June or earlier until September.

Spring budding, although often done in June, may be attempted as soon as the bark slips easily from the stem. Usually this stage does not occur in most species until the sap is flowing well and vigorous growth is beginning. For spring budding, select buds from year-old wood, the previous season's growth. For fall budding, which is actually begun in late summer, use buds grown in the current year. It is usual to collect scion for spring budding before the buds begin to break dormancy, and to keep them in cool storage protected from drying until it is time for use. For summer budding it is best to select the scion ten days or so before it is to be used, clipping the leaf stem about an inch above each bud. Store the bud sticks protected from heat and drying until they are used. This storage period will allow the end of the cut leaf stem to heal over or callus, helping to prevent the entry of disease as well as helping to retard moisture loss after the bud is placed in its

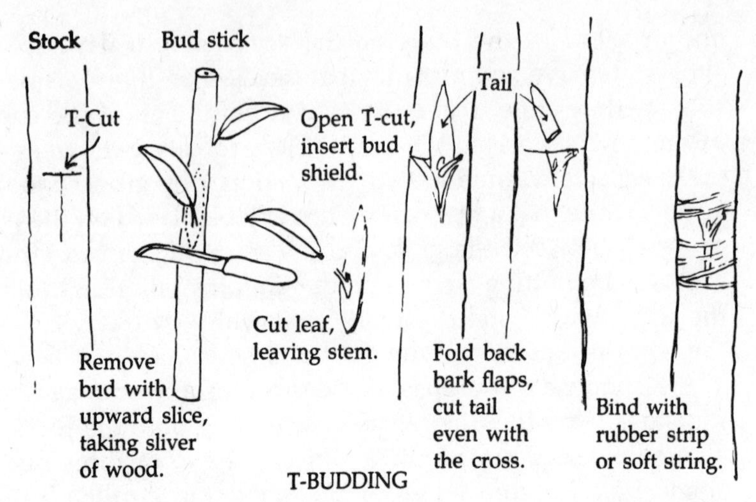

T-BUDDING

new host. The stub of the leaf stem provides a useful handle when you manipulate the tiny bud at the time of the actual budding.

The commonest technique for budding is called shield budding, from the shape of the sliver of bud sliced from the bud stick. The method is also called T-budding, from the shape of the incision made in the bark of the rootstock into which the bud shield is inserted. If you are spring budding, the buds on the bud stick should still be in dormancy; if fall budding, the buds should be of the current year's growth. Otherwise the techniques of budding are identical. The stock upon which the bud is placed should not be more than two or three years old, and it should be sufficiently into the new growing season that the bark separates readily from the stem. On a section of the stem of the rootstock select a section which is smooth and free of branches, about six inches from the ground level. It is not necessary or desirable that the bud be placed where there had been a bud on the stock. Make a vertical cut up the stem about one inch long. Next cross the T by making a horizontal cut tangent to this vertical cut, about one fourth of an inch long. Working from the top, gently loosen the bark at the edges of the cut until they are almost free of the wood.

In a single long smooth stroke of a sharp knife, slice a preselected bud from the bud stick, starting about four times the diameter of the leaf stem below the bud. Slice down into the bud stick and upward, taking a thin sliver of wood with the cut from under the bud, then slope upward above the bud so that the knife comes free, leaving a tail of bark an inch or so long above the bud and the leaf stem. The sliver of wood detached from the bud stick should be carefully and gently dislodged from the bud shield and discarded.

Using both the thin tail above the shield and the stub of stem as handles, work the bottom of the bud shield down through the horizontal cut and into and under the vertical slot—down from the cross of the T down the leg of the T—until the entire bud shield is snugly under the bark. Now remove the upward tail by cutting gently across the top of the T. Only the bud itself and the leaf stem below it should protrude through the slot. The incision may be secured with soft string, rubber strips, or tape. If it is secured by tying, begin at a point below the bottom of the vertical slit and wind upward, taking care when coming up below the leaf stub not to apply pressure that would disturb or strangle the bud, and of course do not cover or disturb the bud from the top. Normally sealants or dressings are not used when budding.

In spring budding, when one is using scion that is dormant, some three or four weeks after the bud has been inserted cut the stock off just above the bud. This pruning will cause the stock to concentrate growth on the single bud, stimulating it to put forth a shoot quickly. Any shoots which the rootstock develops below the bud should be removed as quickly as they are formed.

In summer and fall budding, when one is using scion from the current year's growth, little will occur with the transplanted bud until the following spring, other than that it will remain alive and green. The leaf stem below the bud will most likely drop off. In the fall all ties should be removed to avoid constricting the stock. If there is any curl in the bark at the edges of the incision, the cut should be retied. The following spring while the stock is still dormant or during the normal period for pruning in a given climatic area, the stock should be cut off some four or five inches above the bud to force the bud to develop into a shoot when the growing season begins. This stub, or "snag," as it is called, can be used as a brace the first year to help support and train the new shoot upward as it becomes the trunk of the new tree. At the end of the first year this snag is cut off.

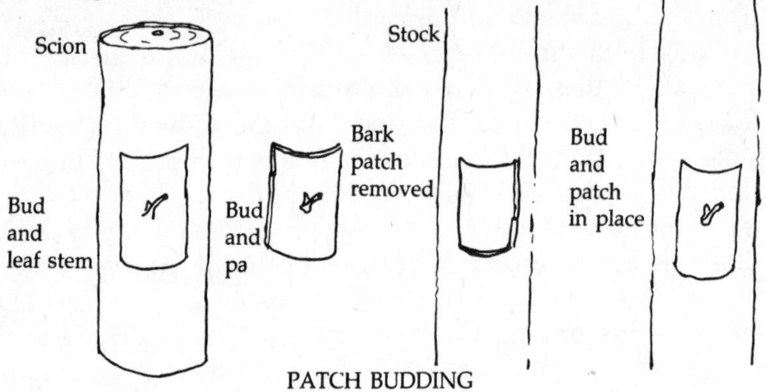

PATCH BUDDING

There is a second pattern for budding that is also useful to know for one interested in propagating his own trees. Although there can be many variations and different names, the basic technique is called patch budding. In this technique a perfect square or rectangle containing the bud is removed from the scion and placed in an exactly matching square or rectangle, left when the bark has been removed from the rootstock. This system is used for propagating trees which have especially thick bark, such as walnuts, pecans, etc. Patch budding is usually done in August or September, using as scion buds from the current year's growth. To make a patch bud requires considerably more exactitude and precision than making a T-bud. The rectangle or square cut from the bud stick must match and fit exactly the recticle cut to receive it on the rootstock. It is possible to make the necessary exactness of fit by working very carefully freehand if one has a knack for precision handwork, but it is far surer and a great deal more easily accomplished by using a special tool. A simple two-bladed knife to make the exactingly accurate cuts can be fabricated out of two single-edged razor blades fastened to a block of soft wood or particle board cut to size. Decide on the size of the patch to be cut, and make a block of exactly that size or a fraction smaller. Attach the two razor blades, top and bottom, with a thumbtack. It is, of course, possible to make a more permanent and rigid mounting, and no doubt worth the time if there are many buds to be set. Most amateurs will seldom be making very many patch buds at a time. There are commercial designs available through grafting-supply specialists.

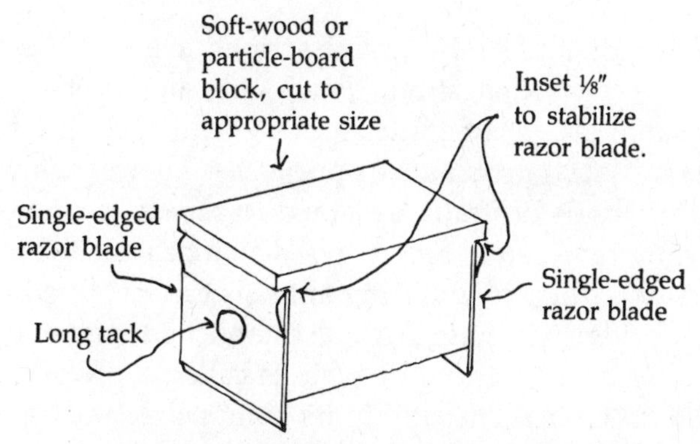

SIMPLE HOMEMADE DOUBLE-BLADED KNIFE
FOR PATCH BUDDING

To use this double-edged knife, position the blades against the rootstock bark, parallel to the ground, on the spot selected to implant the bud. Press, with a slight rocking motion, until the bark has been completely cut through to the wood. This can be judged by feel. Lay aside the tool and, with a sharp knife or another razor blade, connect the two horizontal cuts with two parallel vertical cuts, pressing through the bark to the wood. Work loose the patch and remove it. Repeat the cuts on the bud stick with the double-bladed knife, this time centering the cuts above and below the bud selected for transplanting. Using the bud patch removed from the stock as a template to measure with, again make vertical cuts with a knife or a razor blade, connecting the two horizontal cuts. Carefully, lifting first the edges and then sliding sideways to break the connection between the bark and the wood, remove the patch with its bud. The patch should exactly fit the slot in the bark on the stock and almost exactly match the width. The absolutely crucial points of contact in the fit are the top and the bottom, because the sap will be flowing up and down, not sideways.

Complete the patch bud by tying it with soft string or rubber bands. When making a patch bud it is helpful to use a sealing compound to help retard moisture loss. To serve the same end, the entire patch may be loosely encased in thin polyethylene film. It may also be sealed off with tape. Do not, of course, cover the bud itself with tape or other binding material.

The buds and grafts just described are exclusively for the purposes of propagation. There are, however, many other styles and designs of grafts and buds which are used to accommodate special problems or purposes. There are almost innumerable kinds of grafts designed to repair, to restore or even to totally redesign a tree. A tree which has been completely or partially girdled by gnawing animals can be saved by what is called bridge grafting. An old established tree may be redesigned as a new variety by what is called top working—pruning the tree back virtually to its main trunk and remodeling it with a number of grafts. Weak limbs may be reinforced, even crossbraced, by grafting. To learn how to make such grafts, consult a technical manual. Again I would recommend, both for its completeness and for its clarity, Garner's *Handbook.*

Proper Season and Method of Propagation for Select Fruits

Apple	Graft: whip-and-tongue or splice, winter. Budding: T-bud, spring and fall.
Apricot	Budding: T-bud, spring, summer, fall.
Butternut	Budding: patch bud, summer.
Cherry	Budding: T-bud, summer, fall.
Nectarine	Budding: T-bud, summer, fall.
Peach	Budding: T-bud, summer, fall.
Pear	Graft: whip-and-tongue or splice, winter. Budding: T-bud, fall.
Plum	Budding: T-bud, fall.
Walnut	Budding: patch bud, fall.

NOTE: This chart by no means exhausts the possibilities or combinations. It is simply a suggested list of combinations that work well.

It is even possible, for various practical purposes or merely for the sake of curiosity, to graft among related species of annuals and nonwoody perennials. Cacti are an oddity-maker's delight because of the ease of grafting. But there are interesting possible vegetable combinations as well, most of them more interesting than useful. It is possible to graft a tomato onto a potato, for example, and it will produce, although not very prolifically, both above and below ground. This is not to be confused with the combination tomato-potato offered every year by the less-scrupulous seed catalogues, which is purely a rip-off—at the crudest, a potato with a hole in it packed with a growing medium inoculated with a few tomato seeds. Tomatoes may also be grafted onto the roots of other related species. Joined to the root of the somewhat sinister datura, the plant may produce tomatoes that are hallucinogenic or poisonous. For breeding purposes, peas with disease-susceptible roots are grafted onto resistant rootstock to produce seeds. In the realm of the whimsical, hops may be grafted onto marijuana roots. Grafting combinations are as interesting as they are endless. And with proper facilities and a bit of care and practice, complex as well as playful grafts are not that difficult.

In his concluding remarks in *The Grafter's Handbook,* Mr. Garner quotes from a seventeenth-century Oxonion, Francis Drope: "'. . . there are other variations [of grafting] which I purposely omit, supposing that from these, as from the chief heads, an ingenious lover of his art, will of his own industry, discover and improve them, to his greater pleasure, and content.'"

7. NUTS, AND WHY GROW THEM

Many people back away from planting nuts. Some of this reluctance probably comes from the fact that most of the familiar nut trees tend in size from the large to the huge. Also, some of the most desirable nuts are limited in their range by sensitivity to cold. And, of course, there must be at least a silent concession to human mortality when one's trees may not bear nuts in less than ten years or reach anything approaching maturity in less than thirty. At the extreme, some oak trees will not bear at all until they are past forty years old. That famous seventeenth-century apologist for mortality, Sir Thomas Browne, measured the age of oaks not in years or human lifetimes but in the duration of great families.

All of these concerns are in various ways being addressed by the legions of plant explorers, and the increasing variety which can be found in and through their publications and plant lists greatly eases the strains and restrictions of conventional varieties. Particular interest is expressed by these groups in finding varieties of such nuts as pecans and Carpathian walnuts which have more northerly ranges of hardiness, as well as in discovering or developing larger, thinner-shelled varieties of black walnuts, and such rare but natural crosses as the hican, a cross between the hickory and the pecan. Also, particular attention is being paid to such problems as the blight which virtually exterminated the American chestnut in the last century. There is an organization which is specifically concerned with propagating a few surviving chestnuts that have blight resistance, in hope of eventually reclaiming the tree to cultivation, for the sake of its nuts and fine wood and for its own sake as a severely endangered species. Even so, it is often referred to as extinct, which may hopefully prove a premature judgment.

It is not as easy to be an experimenter and dabbler in breeding and collecting nuts as, for example, it is with apples. While a person interested in apples is dealing with a tree, it is one that can be kept in relatively small cultivation by dwarfing. Even multiple trees are possible with grafting. Some kinds of results are usually possible for the grower in two or three years. Nut trees, on the other hand, are large, sometimes finicky about their pollinization, temperamental about climate, and often require years before they set their first fruits.

But nut trees deserve better than to be dismissed out of hand as obvious impossibilities. While space requirements definitely impose absolute limitations, sometimes, if one views one's property with an eye particularly focusing on the possibilities offered by its perimeters, with sufficient cunning on the gardener's part even large trees may turn out to be possible. Remember, a large tree is using vertical rather than horizontal space. A vining squash or pumpkin, even a healthy watermelon, exploits more horizontal space than a large nut tree. Fence rows or fence corners, odd-shaped patches or even the edges of roads, which may be totally without value for most growing purposes, often can easily support a large nut tree without inconvenience. Perhaps some lesser-known varieties with smaller habits and space requirements should be considered. Some nuts are really no more than bushes and can even be grown as hedges, windbreaks or property dividers or otherwise serve dual functions. And there is always the obvious possibility that is so frequently overlooked: simply to grow nut trees as a regular part of landscaping. For most purposes for which trees are planted, for shade or other yard needs, as for example the so-called "specimen tree," there are nut trees that can probably serve the same function and produce nuts as well. Most of the nut trees are very attractive, and it is hard to find a tree more impressive than a well-grown walnut or pecan.

Nut Trees: General Information*

	Height	Years to Bearing
Almond	12–15 feet	3
Butternut	50–75 feet	2–3
Chinese chestnut	30–40 feet	2
Filbert	12–15 feet	3–5
Pecan	30–40 feet	3–5
English walnut	30–40 feet	5
Carpathian walnut	25–35 feet	3–5
Black walnut	50–75 feet	3–8

* After *Burpee's Nursery Guide.*

Black walnuts will grow virtually anywhere. The chief drawbacks for them as crop trees are the comparatively small size of the nuts, the hardness of the shells and the difficulty of removing the meat after the shell is cracked. Generally the only black walnuts that most people are acquainted with are unselected products of seedlings; basically, the wild black walnut. As trees they will be large and attractive, growing timber that will ultimately be beautiful and valuable. However, if the tree had not been a chance seedling, or a cheap seedling bought from nursery stock, the nuts produced by the tree would have been prized as well as the attractiveness of the tree. One cautionary note should be added about growing walnuts as yard trees. The roots of the black walnut exude a substance called juglone which inhibits the growth of some other plants. Evergreens are said to be particularly susceptible to it. Fallen autumn leaves may kill lawn grass if they are not removed. However, I have never experienced any particular problem with this "walnut poisoning."

Nurseries, particularly the disreputable kinds that promote splashy advertising in mass publications, can be very deceptive about walnuts. In 1981 I saw a full-page ad for what was described as a "flowering walnut." This tree was praised for its lush spring flowers, its nine-month canopy of leaves, and there were many exclamations of "Imagine!" and "Delicious!" Because I had never heard of a flowering walnut before, and because of what I knew of the blossoming habits of other walnut trees, I read the entire wordy message, looking for a clue. Finally, in the small print at the bottom of the page, near the coupon, I found what I was looking for: "*Juglans nigra*"—black walnut. Laws which require listing proper species names somewhat hinder such false advertising, but the real charlatans are hardly slowed down by the requirements. People who may have ordered that "flowering walnut" are still several years away from realizing how badly they have been duped. But at least they have a walnut tree, and that can't be a dead loss.

There are varieties of *J. nigra* which are a great deal more desirable than the common woodlot varieties. Of the named species, no doubt the most familiar is the Thomas black walnut. The appearance of the tree is not particularly different from that of the common varieties, aside from what appear to me to be a greater vigor and quicker growth—although such features may be the result of the greater care and pampering they are likely to get under cultivation. The nuts of the Thomas walnut, however, are unquestionably better in every way. They are commonly one inch, sometimes up to almost two inches, in diameter. They are also thinner-shelled and have a much larger meat. They are still easiest to crack with a hammer, but the meat is not as hard to extract; with a little practice in locating the best place to crack the shell, often the meat may be removed in two large pieces.

Thomas walnuts are offered for sale by most nursery houses, but many of these trees are seedlings, not grafted stock. Seedlings almost never exhibit the fullest potential of the strain. I have two Thomas black walnuts which were planted in 1935. The man who planted them told me that they had been seedlings which a nursery had offered at ten cents a piece. Both are now huge, healthy trees that usually produce abundant crops. But while the nuts are much better than unimproved varieties, they are certainly not as large or as easily shelled as the best varieties of Thomas nuts.

Excellent strains of this walnut may be located through various sources among the guerrilla gardeners. Remember, however, that while most of these people are eager to share and are interested in spreading propagation of their varieties, they do not have commercial resources. Those who seek to make use of what these people have put together should do so with respect, and ideally with reciprocity.

Establishing specimens of the larger varieties of nut trees may not always be easy, but for the most part maintaining them is. Even tender loving care traditionally is unnecessary. The oldest advice about walnut culture I can recall, from the comment of a sixteenth-century scholar about the advantages of using corporal punishment as an aid to education, says that many young boys "like a nut tree must be manuered by beating." Less generously and more generally, similar sentiments are expressed some years later in a rhyme:

> A woman, a dog, and a walnut tree—
> The more you beat them, the better they be.

The picture these bits of lore invite for the imagination suggests someone more in sorrow than in anger flogging the bottom parts of a stolid tree with a cudgel or a strap. What is meant no doubt is a bit more whimsical and less particular. Some kinds of walnut, especially strains of English, Persian and Carpathian, refuse to fall in good time after ripening. In trying to outwait them, I have found them still hanging tight until spring, making the tree look like the permanent roost for numerous small black birds. Traditionally, as today, this habit of clinging overlong may be thwarted by using a long light pole, cudgeling among the branches until the nuts turn loose. It is unlikely that the trees will improve; at least they seem never to remember to give up their fruit the following year and require being beaten all over again.

Walnuts as well as other nuts may be started from seed, grafted or budded from suitable scion wood on one's own rootstock, or obtained as established grafted (usually) nursery stock. Like fruit trees, they may be shipped bare-rooted through the mail or by parcel service. My experience, however, has been that bare-root nuts are less certain of success than bare-rooted fruit trees. Part of the problem is that trees such as walnuts, pecans and their relatives have a very long taproot, a single root that is actually a continuation of the stem straight down from the surface of the soil. A two-foot tree may easily have a three-foot taproot, and the first few inches below ground will probably be larger in diameter than the stem itself. Attached to the taproot will be numerous small feeder roots which, as the tree becomes established, will spread for considerable distances, feeding and anchoring the tree, and may themselves become quite large. The presence of the long taproot is one of the reasons that an established nut tree is so durable. Except in areas where the climate is dry indeed, the root will penetrate the earth to a sufficient depth to find enough moisture to maintain the tree with no further care or attention. In transplanting trees, however, it is difficult to dig up a taproot intact. Bare-root nut trees often arrive with half or more of the taproot cut off. In addition, the few small feeder roots that are attached may be only marginally adequate to support the tree until it becomes established, and when it leafs out the leaves may transpire more water than the roots can replace.

For these nut trees with long taproots, the planting hole should be particularly deep to insure that the soil is loose enough that the taproot may easily penetrate in reestablishing itself. If the soil is especially heavy, mix in a little sand and gravel, and perhaps a very little compost. Do not add manure or chemical fertilizers. Position the tree in the hole to determine how much soil must be backfilled in the bottom to put the tree at its proper height, with the union about three inches above the surface if it is a grafted tree, or at the height it had grown to in the nursery if it is on its own rootstock. After adding the necessary backfill and positioning the tree, sift dirt back in around the taproot, arranging any large lateral roots so that they lie flat and perpendicular to the taproot as the hole fills. Fluff smaller roots so that they likewise will stand out into the soil. Tamp the soil lightly as the hole fills. When the hole is filled to about two inches from the top, add water slowly until the soil is saturated and water stands in the hole level with the surface of the soil. As it soaks up the water, the soil will sink appreciably, so the hole must be refilled until water is again about two inches from the top. The next day again fill the hole with water until the soil is saturated, and add soil as necessary to compensate for the settling. Saturation watering should be continued every other day for a couple of weeks; then gradually extend the periods between waterings until the tree is watered about twice a week. After leaves become well established, it should be watered once a week for the remainder of the summer, more often during periods when the weather is especially dry.

Nut trees frequently will be very stubborn about overcoming the shock of transplanting and breaking their dormancy. Do not give up on a tree if it does not leaf out at all. It is not uncommon for a nut tree to not show leaves until the following spring. I have found pecans to be especially cranky about being moved and slow to restart after transplanting. The first one I attempted, a hardy northern pecan, did not show leaves the first year. The second year it put up a shoot from below the soil level and made a nice growth.

For an attractive shade tree or so-called specimen tree, the black walnut is obviously a good tree to consider in the widest range of climate and soil conditions, especially where hardiness is a prime concern. Another good choice, much better for many people, would be the Carpathian strain of thin-shelled walnut. The nut is essentially identical to the nuts of the common English, or Persian, walnut: the same thin shell, large meat and excellent mild flavor. However, the Carpathian strain is a great deal more hardy than the English walnut. The strain originated in the cold Carpathian Mountains of Eastern Europe and has been grown in North America with increasing popularity since the thirties. The mature tree is smaller than the black walnut, and it has leaves which are a very attractive shade of green, with the individual leaves larger than those of the black walnut and graduated from large leaves at the tip of the leaf stem to smaller leaves toward the base. A mature tree in full leaf gives a pleasing sense of density. These trees also come into bearing more quickly than do black walnuts. Two that I transplanted in 1972 bore their first few nuts in 1974 and have borne increasingly large crops each succeeding year. By their fifth year each was producing about two gallons of hulled (not shelled) nuts annually. In the same period a black walnut would hardly have produced a pocketful. The Carpathian walnut trees are somewhat denser and rounder than black walnuts, and not so tall and upward-reaching. At ten years, my trees are about twenty feet tall, with ten-inch trunks, and their growth seems to be leveling off somewhat. They have experienced below-zero temperatures every winter, and one winter about minus thirty degrees. One exceptionally cold spring both sustained some cold damage, and they did not crop that year.

Most nurseries now offer Carpathian walnuts, usually seedling trees on their own roots. Some, however, offer grafted stock. Also a number of amateurs have collected or developed and are offering—mostly as seed nuts—specific named strains of Carpathians. Collectors are especially concerned with locating and propagating varieties with the greatest hardiness and the largest and best nuts.

While there are named varieties of Carpathians, as yet the development is so new that there is really nothing approaching standardization of strains. Nevertheless, the Carpathian walnut is such a grand tree in every respect that I feel an almost missionary zeal about making it more widely known and more widely propagated. I am constantly passing places where I feel with almost a passion Carpathians should be planted—fence rows, highway dividers, road embankments, etc.

There are other varieties of walnuts as well as black walnuts and Carpathians deserving of attention. The matter is confused by the fact that there is such inprecision in the names of varieties involved. For example, there is one (*J. sieboldiana*) called both locally and widely the heart walnut. This I have also heard called a Japanese walnut, and I long thought they were distinct varieties. Probably the name "Japanese walnut" came from the common practice of giving to any variety of anything distinctly different or exotic the name "Japanese." The heart walnut, at least, seems to have come by the name naturally, from the fact that its somewhat small black-walnut-like fruit has a relatively thin shell which easily cracks to reveal a meat with a distinctly heart-shaped center. The entire nut extracts easily, and the meat is mild-flavored, more reminiscent of the flavor of the English than of the black walnut. They seem to throw true to kind from seed, but are not as hardy as black walnuts. From a nut that my father collected he grew one that fruited after about five years, although the tree was relatively small and slight. Unfortunately this tree winter-killed from an unusually cold winter in what was basically a Zone 6 area. I now have a heart-walnut tree in Zone 5 from a nut collected from a tree growing in Zone 6. It first fruited ten years after planting, but that ten years included at least three years of setbacks from accidents that had nothing to do with species or climate. It has survived without damage winter temperatures of minus thirty. Heart nuts or Japanese walnuts are seldom seen as nursery stock in commercial offerings, but are frequently seen offered by the seed savers. Anyone interested in this variety would probably most easily find it in that way.

Another close kin to the black walnut, and a North American native, is the butternut. The fruit is an elongated hard-shelled nut, deeply indented with numerous sharp-ridged deep creases. The meat is rather thin and difficult to extract. The flavor is mild and rich, and it is easy to see how the name was acquired. The butternut is another variety with which some seed explorers are concerning themselves, being interested especially in obtaining strains that produce larger nuts with thicker meats and thinner shells.

At maturity a butternut tree is considerably smaller than a

mature black-walnut tree. It also comes into production at a much younger age than a black walnut, three or four years as opposed to six years or more. The foliage is especially attractive, with leaves larger and a clearer green than the black walnut, and a smooth, light-colored bark. The tree is quite hardy; although it may not be as hardy as a black walnut, it is usually said it may be grown anyplace a black walnut can. With the selection and attention it is getting from the amateurs, this is a nut tree which might have a promising future. At any rate, it is attractive enough in its own right to deserve consideration for planting as a "yard tree."

Another nut tree which it is hoped may have as important a future as it has had a past is the American chestnut. This tree, which once shaded vast areas of virgin America, was attacked in the last century by a blight which virtually exterminated what was then an important nut and timber tree. The "worminess" of wormy chestnut is almost a criterion of the authenticity of nineteenth-century chestnut furniture. Trees from which this furniture was made had already been killed by blight, and although the dead wood had been infested by worms, because of its intrinsic beauty the timber continued to be used in carpentry as long as any remained to use.

Mostly through the efforts of amateurs and plant explorers, extremely rare surviving specimens of the American chestnut have been preserved and perpetuated, specimens which have survived perhaps because of rare individual genetic resistance to the blight. There is considerable enthusiasm among a small group of amateurs in perpetuating the strain and, it is hoped, futher refining qualities of resistance to the blight. The tree is rarely found commercially, but it is an item which may be located through the seed savers.

In 1982 I planted two seedling American chestnuts which I obtained from one of the few commercial sources. Since I live in a Northwest area far from their native range, I hoped to have an additional advantage to add to genetic resistance, if my climate did not prove to be an insurmountable disadvantage. The seedlings which I received were small, about four inches, but with dense root systems (the American chestnut apparently does not have a taproot, although I don't know this for a fact). My seedlings survived their first season, but it remains to be seen if they can survive longer in the Pacific Northwest. However, this is a species anyone with space and opportunity might well put his hand to as a private experiment which is a potential public service. With many people involving themselves in propagation of this almost extinct species, assuming the heartiness of the surviving bloodlines, eventually this species may be not only precariously preserved as is the whooping crane, it may also actually become reestablished as an important American species in its own right.

A cousin of the chestnut is the chinquapin (which is sometimes called the dwarf chestnut). It has never been, and will probably never be, a nut with any particular commercial potential. The nuts are small, somewhat peg-shaped, with a mild pleasant flavor, growing usually one but sometimes two in a chestnutlike bur. To a certain extent the chinquapin is something of a nostalgic item, reminiscent of schoolboys who collect them and carry them in their pockets for nibbling or trading. Folklore has it that they were (and may be still where the wild nut is still common) an item and object of schoolboy gambling, in an arcane game called "chinkypin."

The nut of the chinquapin might be compared in size and shape to the piñon pine nut, somewhat larger and a little bit more elongated than a kernel of field corn, brown in color, with a thin skin easily cracked and separated between the teeth. The texture of the meat is similar to that of a chestnut, and the flavor slightly more sweet than nutty. In growing habit the chinquapin forms more of a brush than a tree, perhaps comparable to the hazelnut.

Only a few nurseries, and those mostly southern in their location, offer stock of the chinquapin, and I believe the stock offered are seedlings rather than grafted trees. There is some interest in this nut among the seed exchangers, and it will occasionally be seen listed for exchange. With increasing interest in older and native varieties, this nut may be more often seen in the future, more information will be generally available about propagating and growing it, and particular cultivars with especially desirable qualities may emerge. With the stocks presently available, a Zone 6 area would probably be its extreme northern range.

One of the aristocrats of nuts is the pecan. Not only is it a distinctively flavored nut, easy to crack and shell and with a large meat, but the tree itself is especially striking. With adequate space the trees may grow to great heights, tending to form a rather dense canopied top. The leaves are an attractive green, large-lobed, and with an effect that is almost tropical. They are mostly thought of as a southern tree, with all commercial production coming from southern states. Some of the choicest commercial varieties require a 145-day growing season to ripen a crop. Northern gardeners have long hoped to find a pecan that will grow and bear satisfactorily in a much shorter growing season. For many years some commercial nurseries have offered hardy pecans, usually with the advertising being suspiciously earnest about how hardy their strains are. Over the course of time I have tried several in a Zone 5, without ever having one survive more than three or four years. In 1982 I was swayed by an especially earnest catalogue advertisement and was persuaded to try again. If the two trees I ordered and planted do not thrive, I will not be shocked or devastated, but probably will not be cured of falling for another especially convincing claim in the future even if these die.

The people interested in fruit exploring have been active in pursuit of more hardy (and better) varieties of pecans. Their method of proceeding is relatively simple, but demanding. Members and interested parties examine growths of trees, especially remote groves, for pecan trees in places where they don't really expect to find them. They investigate waterways especially, and they have found specimens in the upper Mississippi system a great deal farther north than native trees had been previously known. Curiously, and as an indication of their thoroughness and zeal, after receiving assurances from the natives of a locale that no pecans have ever been seen in that area, frequently the fruit explorers will find a tree.

Results developed by the tree explorers are perhaps more promising than cultivars offered by commercial sources. Nuts have been found in quite northern regions, and members sometimes offer varieties they have found for testing and reproduction. In a few years it may be that the term "hardy pecan" will be honest and accurate.

Hazelnuts and filberts are another good nut for a gardener's consideration. Essentially the two names designate the same nut, although "hazelnut" usually implies a native or wild variety, while "filbert" implies an improved cultivated strain. In England the hazelnut is also called a "cob nut." Hazelnut trees are not really trees at all—more like disorganized tangled shrubs which grow to a height of about fifteen feet. In fact, the country idiom usually refers to "hazel brush" rather than to hazel bushes or hazel trees. Because of these habits of growth, this nut family can be used as a hedge, as a windbreak or incorporated into roadside areas or other "outside" locations.

The named cultivars of filberts, the kinds most usually seen offered by nursery houses, are somewhat more sensitive to cold than the wild varieties. As with pecans, catalogues tend seductively to suggest that their strains are especially hardy in northern areas. I have never been able to get any of the named filberts to survive in a Zone 5 area, although they can be grown in a Zone 6 and are commonly grown commercially in the most temperate parts of this zone.

However, I have had very good experience growing wild hazelnuts. I have four bushes grown as a hedge-windbreak-divider along a back fence, and although they were badly damaged by the cold one extremely hard winter, they came back rapidly from their roots and were soon producing as well as ever. The nuts are smaller than those of the commercially named filberts, but their small size causes only a little more trouble in shelling than the larger varieties. They are quite good, but whether they are appreciably or even any better than commercial filberts would be hard to attest to. Those four bushes produce as many nuts as our family can use, and if they were larger they would certainly produce considerably more than we could use. Someone interested in trying the wild nuts should have no trouble finding seed exchangers willing to swap them and other cultivars of the family as well.

All of the nuts discussed have both male and female blossoms and are pollinized mainly by the wind. All varieties of nuts—walnuts, pecans, hazelnuts—produce much more prolifically if more than one tree is grown in proximity. Filberts are more fussy. Because the pollen-bearing male blossoms and the fruit-producing female blossoms do not mature at the same time, most varieties are self-sterile. Therefore not just two different plants but plants of two different varieties must be cultivated in order for them to fruit. Wild hazelnuts, however, pollinize mutually with no problem, which may be a convincing argument for planting the wild rather than the cultivated varieties.

8. BERRIES AND UNCOMMON NATIVE FRUITS

The grandest compliment the strawberry has ever received is William Butler's famous observation that God certainly could have made a better fruit than the strawberry if He had chosen, but that it is equally certain that He did not so choose. The genial Butler would no doubt have been more bemused than angered at the efforts mortal plant breeders have expended in the last century and a half attempting to do what God did not choose to do, improve the strawberry. Among the cornucopia of goodies the New World has sent to the Old there has been one bad apple: the South American strawberry.

There had been two consistent complaints about the European strawberries, neither of which had anything whatever to do with flavor: the berries were small, and their period of production was narrow. The limited genetic pool of European strawberries had not been sufficient for early plant breeders to make any particular breakthroughs in improving on these limitations. The New World offered new gene pools. Several varieties of South American strawberries that had nothing to recommend them but their size were taken to Europe. One was white, and in consistency and flavor it was more like a willow cone than a berry. Exactly what happened in the course of breeding in Europe is uncertain. Eventually, however, the prototype of the modern strawberry appeared, the result of various crosses with these South American imports and European and North American wild strawberries. Few would agree that man has succeeded in doing what God did not choose to do, make a better fruit than the wild strawberry.

Very likely the strawberry praised by William Butler was either the *Fragaria moschata,* popularly called the hautbois or mush strawberry, which was an established variety in Europe by the sixteenth century, or the *F. vesca,* which was probably grown in Rome as early as 200 B.C. The *F. moschata* is reported to have both unique flavor and aroma. Presumably it continues to be grown in Europe, but I have never found an offering of it in American catalogues or plant lists. I am confident, however, that somewhere in the underground of the guerrillas it is to be found. The *F. vesca* continued to be generally popular in England and France until the nineteenth century and continues to be the gourmet French strawberry.

The usual name of the *F. vesca* in France is *fraise du bois.* Some commercial catalogues have been offering it more or less as a novelty item under the name of "Alpine strawberry." The berry is somewhat conical and elongated, and quite deep red in color. Because of its numerous and prominent dark seeds the berry has an almost freckled appearance. The flavor is distinctively sweet-tart, with an unmistakable full-bodied wild-strawberry richness. The plant is compact, with leaves that are somewhat narrower than those of most domestic strawberries. Each plant is quite productive, although the individual berries are rather small. A few nurseries offer plants, but the prices tend to be rather steep, probably because the plants do not propagate by runners. Other commercial houses offer seeds; since these are nonhybrid, open-pollinated varieties, seeds will throw true to the parent stock. If the seeds are planted indoors in late winter, they may be transplanted outside as the weather begins to warm, and they will produce a respectable crop the first year.

Strawberries of all kinds are rather easy to grow, and although there are varieties more suited to the South, and varieties more suited to the North, strawberries in general do well under a wide variety of climates. Nor does it require an exceptional soil for them to do well. It is frequently recommended that they be planted in ground that was under cultivation the previous year, as this gives the soil a good loosened texture.

Crops will be improved if the soil has been enriched with manure or other organic material, well worked in the previous year. The bed should not be fertilized before planting, nor anytime during the growing period of the first year. Fertilizing during these periods tends to encourage rank growth of the plants at the expense of fruit, and it seems also to result in the berries being overly soft and to not keep well on the vines.

In the northern states strawberries are usually set out in April or May; in the central states in March and April; and in February and March in the southern states. On the other hand, some gardeners plant in the fall, and this practice seems to be gaining acceptance. In 1981 I tried this for the first time, making transplants from runners of my own plants in early September. A small percent winter-killed, but most survived. They put on a good spring growth and had a respectable if light crop of berries.

When one buys plants from a commercial catalogue, the buyer usually is asked to indicate a preferred time for delivery. If when the plants arrive it is not possible to set them out immediately, they should be kept under refrigeration in their shipping bags. An ideal temperature for keeping them is in the middle range of thirty degreees. They may be kept for several days under refrigeration, but there is an urgency to get them into the ground quickly. Just before transplanting, the roots should be soaked in water for half an hour or so. Soaking is not imperative, but it will give them a good boost in breaking dormancy and becoming established.

Just before setting the plants into their beds, the soil should be retilled and all clods broken up. Set the plants about eighteen to twenty inches apart. If more than one row is being planted, space them about three feet apart. Commercial plantings would space them farther. Remove the plants from their binding—usually a rubber band—and separate the individual plants from the bunch in which they are packed. It is important that the plants be kept moist until they are set into the soil. Even if the soil has been properly loosened, I usually dig a hole six or eight inches wide and six inches deep with a trowel, sifting the excess soil back into the hole where the plant will be set to assure the roots an unobstructed medium in which to grow. It is very important to set the plant into its hole at just the proper depth, not too deep and not too shallow. The optimum to be achieved is to have the plant reset at exactly the same level it was growing before it was harvested. Therefore place each plant in the soil to a level just above where the roots join the crown—the fleshy stem from which both the roots and the leaves are growing. If any roots are showing when the plant has been placed in the ground, it is too shallow; the plant will be in danger of dehydration, which may kill it or, if it survives, cause it to be stunted. If it is set too deeply, with dirt coming up around or on top of the crown, very likely the plant will rot.

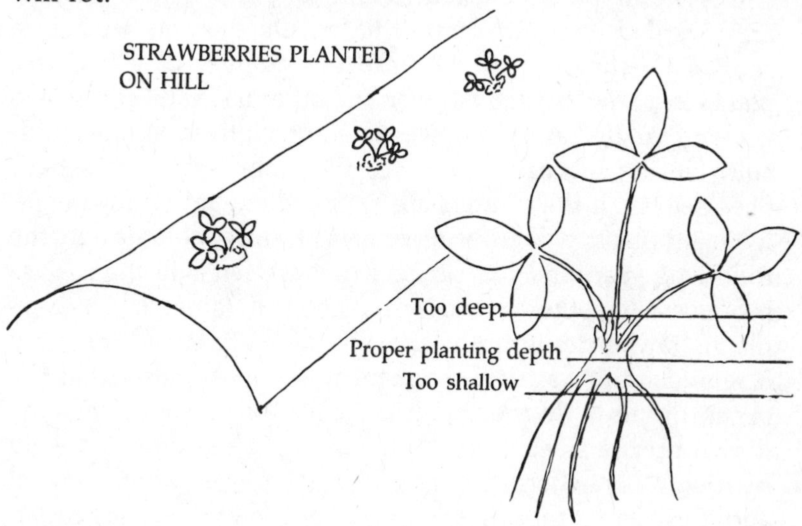

When the plants have all been set into the soil at their proper depth, each should be carefully watered; give enough water to saturate the soil, but not enough to leave standing puddles. Be careul, of course, not to dislodge the new plants while watering. If there is not sufficient rain, irrigation should be provided frequently enough that the soil remains moist until the plants are fully established, which takes from two to three weeks. Most likely about a month after the new plants have been set out they will begin to form flower buds. The plants will grow much stronger and eventually produce much better crops if these first buds are picked as they appear and not allowed to develop into fruit.

Most strawberries grow runners, although some varieties grow few or none. A runner begins as a stem which grows from the central crown. On its tip is a bud, and when the stem is about twelve inches long the bud develops a rosette of leaves on its top and from its bottom puts down roots. Often additional new runners will develop from this rosette, putting up leaves and setting down new roots, and so on. Each of these rooted rosettes will form a new strawberry plant, and this is the commonest way most varieties are propagated.

Depending on circumstances and conditions, these runners are regarded with different attitudes. On the one hand, in a short time runners will increase the number of strawberry plants in a row enormously. On the other hand, these runners get very tangled and make weeding and cultivation most difficult. During the first year after the strawberry plants have been set out, if the runners are kept pruned until midsummer the main plants will grow larger and produce more heavily the following year, and the runners that are left will themselves produce some the second year. Many commercial growers plant strawberries in rows each consisting of a steeply elevated hill, which allows for convenient mechanical cultivation between the rows. In the process of cultivation, the runners are drawn up the sides of the hill, forming a more or less solid matting of strawberry plants, and making use of both vertical and horizontal space. It also makes picking the berries easier. In practice, however, it doesn't seem to work out with the uncomplicated simplicity the design seems to promise.

Strawberry plants do not have a very long productive life. Their production will peak in their second year and begin to decline in subsequent years. After five years the vines may continue to look fine but produce very few berries. Some gardeners regularly rotate their strawberries on a two- or three-year time scheme, planning and making advanced preparation for the next bed at the same time a new bed is being planted. Once one has an established strawberry bed, plants started from the runners can be used to make the new planting. In fact, these runners will do much better than ones obtained as nursery stock. One system of rotation involves preparing the new bed in the spring as a three- or four-foot band paralleling the established bed, adding manure or compost and tilling the soil thoroughly. In the middle or late summer, runners are trained from the established plants onto the newly prepared strip. The second year the runners will develop into maturity, and after the crop from the old berries is finished that bed is tilled under. If one chooses, he can leapfrog rows of berries in this fashion back and forth across a garden plot indefinitely without ever actually transplanting any plants.

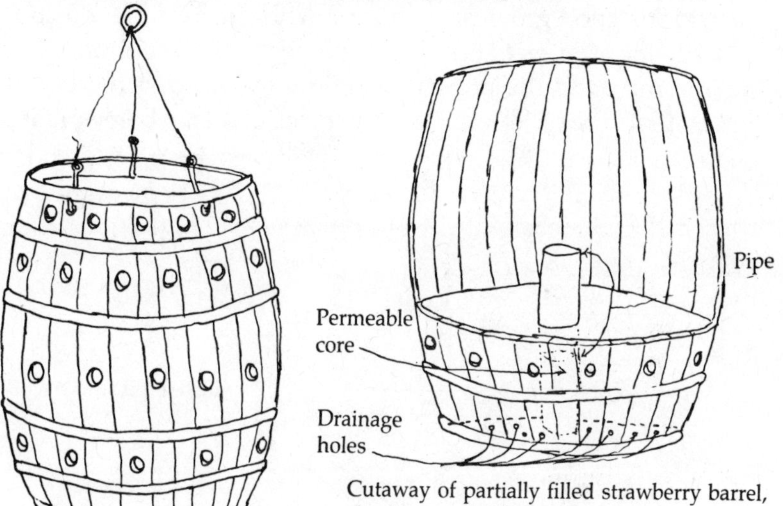

Strawberry barrel ready for planting

Cutaway of partially filled strawberry barrel, showing method of using a pipe to provide a permeable core through the center of the growing medium

A traditional and picturesque way of growing strawberries that is both effective and space-saving is to use what is called a "strawberry barrel." Into the sides of an old fifty-gallon wooden barrel (smaller barrels as well as steel barrels are also used) holes about an inch and a half in diameter are bored, spaced about ten inches apart. The holes may be positioned in patterns, randomly, or in circles paralleling each other around the circumference of the barrel. For drainage, eight or ten smaller holes should be bored in the bottom of the barrel. The barrel may be suspended from a beam or a scaffold, placed on bricks or cement blocks, or put on a mobile platform. However it is to be mounted, the barrel should be in the position it is to occupy before the growing medium is added. Once filled, it is very heavy.

The growing medium needs to be rich, because it will be supporting a heavy population of plants. It must also be of a texture that allows easy penetration by roots and free movement of moisture. A good soil mixed with compost or manure, to which has been added about twenty percent peat moss, provides a good growing medium. With the barrel in position, put a layer of gravel an inch or two deep on the bottom of the barrel, then add the growing medium to just past the bottom of the first row of holes. Place the roots of a stawberry plant in

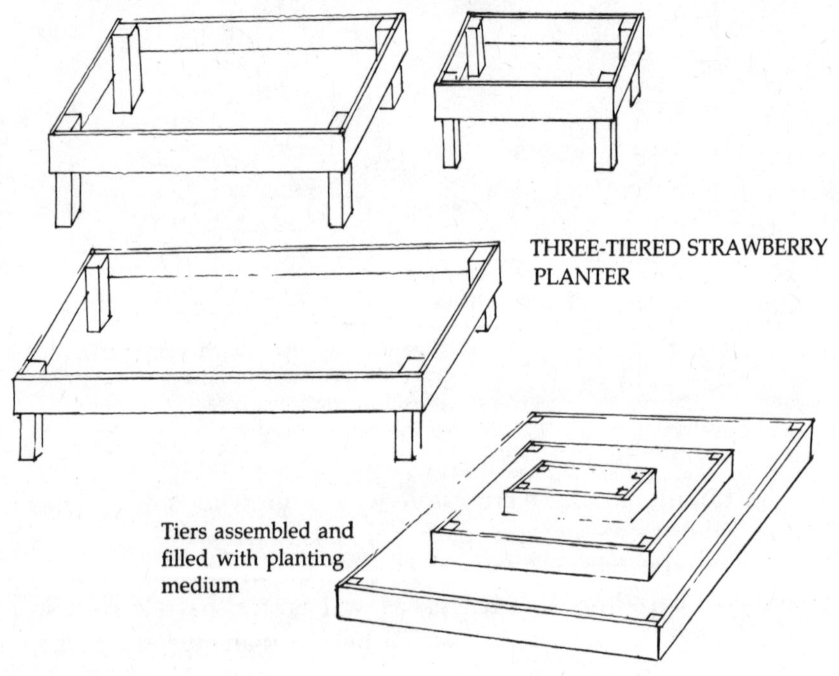

THREE-TIERED STRAWBERRY PLANTER

each of the holes, with the crown outside the barrel, packing soil around it from the inside so that it is neither too deep nor too shallow, exactly as if the plants were being planted in the normal perpendicular fashion in the garden.

At this point a watering system may be incorporated into the strawberry barrel. Even with a growing medium that allows easy passage of water, it may be difficult to keep the soil at the bottom moist without drowning out the top. This problem may be overcome by preparing a core of highly permeable material down the length of the barrel. Such a core may be provided by packing the material into a tube placed in the center as the growing medium is being built up. A pipe made of light thin metal such as stovepipe or gutter pipe is ideal. It need not be long; the length of a single joint of stovepipe is sufficient. Place the pipe upright in the center of the barrel resting on the planting mixture, and fill the pipe with a fifty-fifty mixture of sand and peat moss to the top. Add more planting mixture up to the next hole or layer of holes, and

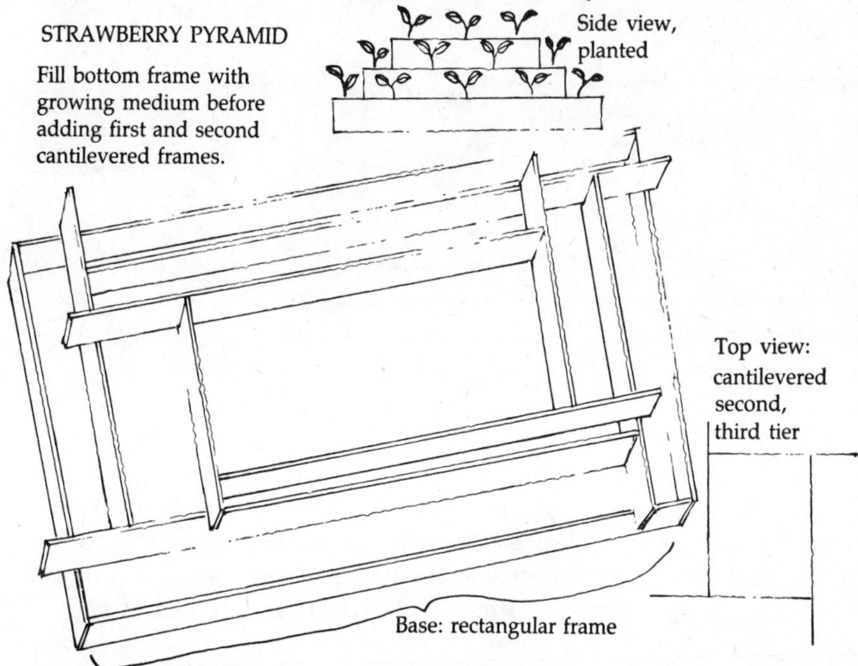

plant strawberries in each as before. When the level of the soil mixture reaches about two-thirds the length of the outside of the pipe, carefully elevate it, making sure the sand-and-peat-moss mixture inside remains in position to provide a core down through the center of the planting mixture. Lift the pipe until six inches or so remains under the soil for support. Again refill it with the mixture of sand and peat moss. Proceed in this way until the barrel is full of soil, and berries have been planted in each of the holes. The pipe is now removed completely, leaving an uninterrupted core of permeable material down through the center of the barrel to serve as a water conduit. Gently add water to this core until the planting mixture is thoroughly moistened.

The strawberry barrel should be watered frequently enough that it will never dry out in the slightest. The plants should grow quickly, and by summer's end foliage should completely cover the barrel. Particularly when it is covered with blossoms and berries, a strawberry barrel is very attractive and provides a fine accent point for a garden, patio, etc.

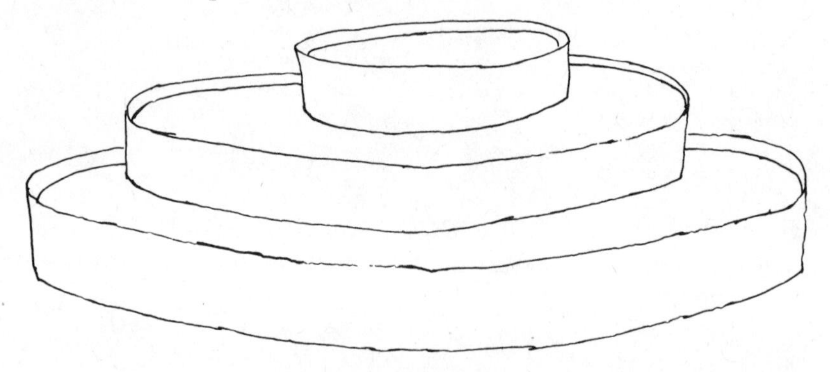

Strawberry Pyramid Made from
Metal Strips Fastened in Circles

A comparable space-saving arrangement making use of vertical rather than horizontal space is called a strawberry pyramid. Several different arrangements of concentric aluminum rings, etc., are commercially available, but with the general plan or pattern in mind attractive homemade designs are not difficult or expensive to improvise. The accompanying illustration is self-explanatory and should suggest other designs which can be adapted to available space, building materials, construction skills, etc.

Alpine strawberries look especially attractive grown in a barrel or a pyramid, because of their rather bunched configuration and unusually shaped foliage as well as the particular visibility of their blossoms and berries. Such growing arrangements also provide appropriate showcases for rare and unusual varieties of strawberries that one might gain possession of.

Although, to my knowledge, little has turned up in the way of heirloom strawberries, I suspect that they might become an item of interest. Not only are there probably European varieties that might be recovered for American gardeners, but there are numerous North American wild varieties to be explored also. A cultivar of Virginia wild strawberries was widely grown in America in the eighteenth century and exported to Europe as well. It is probably one of the berries that were crossed with the pithy South American varieties in producing the modern commercial strains. Perhaps descendants of the original cultivars still exist here or abroad, or perhaps the wild ancestors will be rediscovered for domestic cultivation.

In the past I have tried with no success to transplant wild strawberries from the mountains of the Pacific Northwest to gardens in various lowland locations. Sometimes the plants would not grow at all, and even when they actually thrived they would never fruit for me. Differences in climate, soil, etc., between their native habitat and mine were the probable culprits. However, with the numerous potential varieties the United States and Canada have to choose among, eventually amateurs will certainly collect and then disseminate specimens until all who choose may have berries as God made them and Walton ate them.

Among the so-called caning and vining berries—such as raspberries and blackberries respectively—the advantages of seeking heirloom varieties are considerably fewer than is true with strawberries. So far these berries for various reasons have not been subjected to heavy commercialization. They do not adapt well to mechanical harvesting and are inherently so fragile as to make their distribution by definition local, and long shipments impractical if not impossible. These limitations make such berries especially suitable for home-garden production. Except locally, and then mostly infrequently, raspberries, loganberries, dewberries, boysenberries, blackberries, etc., will not be commercially available. Except for adaptability to different climates, the varieties offered by most commercial nursery houses are fairly similar. Unlike apples and pears, for example, the varieties found commercially are pretty much the only varieties available anywhere. Some uncommon varieties are offered for exchange among the seed-saving clan, but mostly even they reflect the varieties commercially available. I have been generally quite satisfied with varieties I have obtained from the commercial sources.

Before ordering cane plants from a nursery, always check around the countryside for starts. Even though there is no data sheet on varieties, you may find some of proven local quality, and such local transplants will always be better than mail-order specimens. And, of course, cheaper. I have some very excellent berries obtained through mail-order sources, but my favorite raspberry is an unnamed variety a friend gave me.

I have not seen much attention paid yet to blueberries among the exchanges, but this could be another area that simply has not yet awakened. Blueberries are a commercially valuable crop, and much more than raspberries they are equal to shipment and a certain degree of handling. However, I have never tasted a commercial blueberry that was worth comparing to the wild huckleberry. Wild huckleberries grow in many places across the country, and it must be possible that some of these berries can be adapted to garden cultivation. True, this family of berries has fairly restricting soil requirements, but these soil conditions can be met in the garden by taking a little trouble. While I have tried transplanting wild huckleberries several times, I have never had any luck in establishing them. As in my experience with wild strawberries, probably the difference between where they were growing and where I transplanted them was too extreme. Wild huckleberries should be a subject for experimentation.

If one is interested in growing them at all, gooseberries are worth a little experimentation. The name, by the way, has always been something of a linguistic enigma. Geese do not eat them. To test this, I once cast gooseberries before my geese and they were absolutely ignored, and they were about the only edible thing I ever saw geese ignore. Thinking that perhaps it was simply the naïveté of my rustic geese that made them fail to recognize a good thing when they saw it, I caught the geese one by one and forced berries into their beaks. The geese spit them out with indignation. The name of the berries may have caused me to be overly insistent, but still things like that should be checked out. The name has also been explained as a corruption of *gorse* berry, but the shrub in question certainly is not a gorse. A more likely and (to me) convincing explanation is that the name comes from the French *groseille à maquereau,* translated roughly as "mackerel currant," from the use of this berry with fish dishes.

Except regionally, gooseberries enjoy little favor in America. They are quite a popular fruit in Europe, however, especially in England. The first time I was in London and poking through the produce markets—"fruiterers," in the British vernacular—the most striking difference I noticed between them and their American counterparts was the number and varieties of gooseberries offered. In fact, I think I have never seen gooseberries in an American market. In each British market I found larger berries than in the last. Some were fully an inch in diameter, and in color they varied from the familiar gooseberry green to canary yellow, strawberry pink, tangerine orange and unspecifiable shades besides.

Most likely a major reason so many varieties exist today in Europe is a curious fad of the eighteenth century. For some reason in the late eighteen-hundreds and early nineteen-hundreds a passion for breeding gooseberries developed among fashionable young Englishmen. There were gooseberry clubs, gooseberry shows and gooseberry fairs. By 1831, it is reported, 722 cultivars had been named. The young rakehells of the Canterbury Gooseberry Club would be devastated to see a modern American catalogue listing at most five varieties but commonly no more than one.

The gooseberry would be an ideal plant for a neophite plant collector to dabble with. Scrutiny of the various exchange lists should turn up interesting and oddball variants, although not likely all of the 722 varieties listed in 1831. Gooseberry bushes are extremely easy to grow, and the common varieties at least are exceptionally hardy. An established plant will stand up to endless neglect and even abuse. I have one plant which grows in a most desolate fence corner overshadowed by a hedge of hazelnuts, out of sight and far out of mind. It never gets water in the summer other than our niggardly rainfall, yet it produces probably two gallons of berries each year, if they ever got picked.

Most gooseberries are uncomfortably thorny, and picking them can be a bloody business. Varieties which are thornless, or nearly so, have been developed, as well as a variety which offers its berries in such an arrangement as to make them especially easy to pick. However, if one is intentionally going to plant a gooseberry, he might just as well try one which is grown for the value of the berries themselves. Many people find that gooseberries have little enough going for them in their own right anyway. But perhaps they really are a treat with mackerel if the right variety is found.

An unusual American fruit which is currently of particular interest to the fruit explorers and other guerrilla gardeners is the pawpaw. It was once a quite common and widely distributed wild fruit, sometimes brought under cultivation, but it is becoming increasingly rare as its wild native habitat diminishes. My mother, who remembered it fondly from her girlhood in southwestern Missouri, on a nostalgic visit to her birthplace was unable to locate a single plant.

The pawpaw is a close relative of the papaya, and its tropical-looking foliage and fruit both seem wildly out of place in a temperate climate. The fruit almost never is found in commercial markets—in my experience, never—but it has a most respectable reputation as a real delicacy when it is grown locally. Considering its tropical associations, it is surprisingly hardy. Some of the few nurseries that offer plants for sale overdo this reputation for hardiness perhaps, especially in referring to it as the "Michigan banana" and calling attention to the fact that there is a town in Michigan named "Pawpaw." In spite of extreme claims, pawpaws, or at least most strains of them, do not survive harsh cold.

Twice I have tried to grow stock which I obtained commercially, having been tricked again by catalogue hype during the annual period of winter gullibility. In both cases the plant that arrived had been a poor thing to begin with and was abominably packed, or rather not packed at all. Neither plant ever sprouted a leaf, despite the tenderest and most solicitous of care. Certainly pawpaws would be very chancy in my Zone 5 climate, but I would like to try them anyway. I do know they were grown fifty miles away in the microclimate of the Snake River in the early part of this century.

The fruit explorers' organizations are devoting attention both to collecting and to propagating pawpaws. They have extensively examined old plantings as well as wild growths in as many parts of the country where they grow. Aside from generally expanding knowledge about the range, quality and varieties, they are interested in repopularizing the fruit. They are reported so far to have found large variations among cultivars: varieties with few seeds, varieties that excell in flavor, varieties with exceptional hardiness, etc. It would be a safe guess that through the efforts of these fruit explorers the pawpaw will become much better known in the future and more varieties will be appearing among the listings in the fruit exchanges. Perhaps something will turn up hardy enough for my

Zone 5 garden. It seems, by the way, that pawpaws are especially difficult to transplant and are most easily propagated from seeds. I have read nothing about how well they throw true to parentage from seed.

Mulberries are not a particularly uncommon fruit, but except in certain locations they are not widely known in cultivation. The fruit of the mulberry resembles a small raspberry, varying in color from yellow through red to dark, dull black. There are native American varieties and Oriental varieties, as well as crosses between them. The Oriental variety has been grown primarily for its leaves, which are fed to silkworms, and was introduced into the Western world when silkworms were first pirated from the Orient. These trees are of medium size, commonly perhaps twenty feet (although I have heard of much larger specimens), with a round, rather symmetrical top. There is much variation in the flavor of the berries. Those I have eaten have tasted rather blah, just a degree away from being unpleasant. Desperate small boys will eat a few of them. However, my personal experience with mulberries has been limited, certainly with no more than half a dozen trees. Someone else may have been getting all the good ones.

This is another variety of tree that some fruit explorers are actively interested in. From articles I have read in the literature of the underground, there seem to be many better varieties than I have had contact with. While note is taken of varieties with insipid flavor and flaccid texture, other varieties with good to excellent flavor as well as firm texture are reported on. Mulberry fanciers are also working with crosses directed to producing or improving specific qualities.

Sometimes in gardening literature one runs across the advice that mulberries will attract birds away from cherries and other vulnerable fruit. At least one tree I have encountered was planted for that purpose, and its presence did seem to help. There is also the possibility that it was attracting birds which would then move on to the cherry trees when the mulberries gave out.

In some areas of Arizona mulberries have been extensively used in recent years for hedges and landscaping, particularly around newly constructed retirement areas. Ironically, people who have moved to Arizona for its dry and pollen-free air are finding themselves thwarted by ubiquitous mulberry pollen. This, then, might be a consideration to be taken into account before planting mulberries: many people seem sensitive to their pollen.

Another tree, partly native and partly introduced, which the fruit explorers are collecting, testing, and propagating, is the persimmon. Although this fruit is frequently seen in markets in late autumn, it is still relatively unfamiliar to many people, especially in the northern states. Familiar or not, it is a part of our folk vocabularies as a synonym for "bitter." Bad press may have contributed to its lack of popularity. The native American varieties are puckery and inedible until very late fall, and they require a frost before they become sweet. Usually most of the leaves will have fallen from the trees before the fruit is picked. A leafless persimmon tree, loaded with fruit, is a strange and exotic sight: with the bright-orange to orange-red blunt-cone-shaped fruit, the trees resemble something decorated for Christmas by a very single-minded Santa Claus.

The flesh of a properly ripened persimmon is soft and pulpy, with a mild sweet flavor perhaps a little like that of an apricot. Those who have eaten superior varieties might well modify this description and lace it with more enthusiasm. The persimmon tree is not especially hardy, although those people who are experimenting with the species may cause changes in such judgment. The collecting, testing and breeding by the fruit explorers is likely to have considerable effect on the popularity and cultivation of persimmons in the near future.

APPENDIX I
COMMERCIAL CATALOGUES

The following list of commercial catalogues is not complete, and probably it would be impossible to compile a list that was truly complete. Even though they basically exchange seeds and plants among themselves, many of the plant collectors and connoisseurs of rare and unusual plants make extensive use of these commercial sources as well as of underground sources. In addition to providing seed and stock for planting, such catalogues are extremely useful for the information they contain about the cultivation, use, and sometimes the origins and backgrounds of plants. While those on this list are referred to as "commercial," there is a fine distinction, which is difficult to make, between what is and what is not technically commercial. The publication *Southmeadow Fruit Gardens,* for example, is a comprehensive and descriptive book. Accompanying it, however, is a price list for ordering the varieties described in the book. I list it here as commercial because it provides an avenue through which varieties of rare and heirloom fruit trees may be obtained. The compilation itself has to be considered a labor of love. A distinction in such a case is an arbitrary one.

I have examined almost all of the catalogues listed, although I have listed some catalogues which I ordered but did not receive. In a few cases, I have indicated the importance of a given catalogue in the notation although I have not specifically seen that particular catalogue, because others have called it important. Commentary on the vast majority which I have seen is necessarily sketchy and uneven. Perfectly fine catalogues have received perfunctory treatment if they did not happen to have varieties that I thought might be interesting to a fancier of old or unusual varieties. Such judgments have to be arbitrary and even capricious. One man's oddity is another man's commonplace.

The listings and the commentary do not constitute endorsement. I have ordered from many of these catalogues, and I have indicated my experience with them (overwhelmingly positive), but this does not imply endorsement. In transactions with the catalogues there was never an indication that those which I was ordering from might be included in a listing such as this. In many instances my experience with the companies goes back for decades. Many are so small or private or just plain independent that they wouldn't care in the slightest for what anybody said about them anyway.

Where a price for a catalogue is expressed or known, I have listed it. For some catalogues this price is refundable with the first order. Even if a price is not stated, for the small companies it would be a kindly gesture to include two or three first-class stamps with a request for a catalogue. If the small independent companies are to continue to survive and provide the service and quirky selections that only the independents can offer, a few stamps are not only a courtesy, they are an investment.

ABC Nursery and Greenhouse
Lecoma, MO 65540

Their leaflet is only a partial listing of their stock and is mostly devoted to herbs.

Adams County Nursery, Inc.
Aspers, PA 17304

This small catalogue lists apples, pears, peaches, apricots and cherries, including many old varieties. It is a very professional nursery, listing a variety of rootstocks. The trees I have ordered from them arrived in first-class condition and were excellent specimens. Their small catalogue also gives brief but useful information about handling and planting.

Agway Inc.
P. O. Box 487
Elizabethtown, PA 17022
The seed varieties listed are fairly usual, but the information about cultivation and use of each variety is unusually full and readable. As a centerfold they have an exellent planting guide.

Ahrens Strawberry Nursery
R. R. 1
Huntingburg, IN 47542
The catalogue includes a wide variety of berries besides strawberries. It is an attractive catalogue, and has a great deal of information about cultivation of berries.

W. F. Allen and Co.
P. O. Box 1577
Salisbury, MD 21801
While they specialize in strawberries, they offer other berries as well (although no heirloom varieties). The catalogue is attractive and informative.

Alston Seed Growers
Littleton, NC 27850
This company offers many varieties of heirloom field corn. It is a very good and useful source.

Altman's Specialty Plants
553 Buena Creek Rd.
San Marcos, CA 92069
This catalogue is exclusively devoted to succulents. Its offerings are extensive, and it is well illustrated and detailed in explanations.

Applewood Seed Company
P. O. Box 10761
Golden, CO 80401
This company specializes in wildflower seeds, but also offers plants for use in dried arrangements, as well as houseplants.

The Banana Tree
715 Northampton St.
Easton, PA 18042

Most of the trees and plants listed are tropicals, although they list some that are cold-hardy. Many varieties are rare or uncommon. They do list a wide variety of banana trees.

Vernon Barnes and Son Nursery
P. O. Box 250 L
McMinnville, TN 37110

This small catalogue offers ornamentals, fruit trees and berries, and quite a few nuts. The selection is small but useful, and the prices exceptionally low.

Baum's Nursery
R. D. 4
New Fairfield, CT 06810

This catalogue lists a number of old varieties of apples.

Beersheba Wildflower Gardens
Beersheba Springs, TN 37305

This is a small, very attractive catalogue of native plants (not seeds).

Boston Mountain Nurseries
Rte. 2, Hwy. 71
Winslow, AR 72959

This nursery specializes in berries of all kinds. It is a good source for cane berries.

Bountiful Ridge
120 Nursery Lane
Princess Anne, MD 21853

Fruits, berries and nuts are specialties. A very large listing, including rare and heirloom varieties. It contains a great deal of useful information about varieties as well as about culture, rootstock, etc. It is well worth owning.

Brittingham Plant Farms
Dept. 0-6
P. O. Box 2538
Salisbury, MO 21801

This catalogue lists berries of all kinds and seems to cater to the small commercial trade. It includes old standard varieties of berries, but not varieties that are particularly designated heirloom.

Broom Seed Company, Inc.
P. O. Box 236-N
Rion, SC 29132

This small catalogue offers a mostly conventional list of varieties of seeds, but specializes in small packets at a correspondingly low price. It would be a very economical place for someone with small needs to shop.

John Brudy Exotics
P. O. Box 1348
Coca Beach, FL 32931

This catalogue offers a fascinating variety of seeds that make one wish he lived in a warmer climate. It offers seeds from such diverse and unusual plants as the balsa tree, cashew, cacao trees, palms of all sorts, papaya, etc.

W. Atlee Burpee Co.
Warminster, PA 18974

This is a standard nursery of high quality. It lists mostly standard hybrid varieties. It publishes a very useful booklet on plant culture.

D. V. Burrell Seed Growers
Rocky Ford, CO 81067

This company offers a list of fourteen varieties of peppers. It also has an especially large offering of melons.

California Nursery Co.
Box 2278
Fremont, CA 94536

This nursery sells a large variety of nursery stock, especially varieties suited to warm climates, such as olives, pomegranates, kiwi fruit, persimmons, etc. It has a good selection of less-common grape varieties, including wine grapes.

Casa Yerba
Star Rte. 2, Box 21
Days Creek, OR 97429

This catalogue offers a quite large range of herbs and sells some plants as well as seeds. It provides brief comments on the cultivation of each variety.

Cloud Mountain Farm
6906 Goodwin Rd.
Everson, WA 98247

The selections are not large, but they are good. They offer some old varieties of pear, cherry, plum and apple, particularly selected for adaptability to marine climate. They offer a few varieties of wine grape. They are one of the few catalogues that offer named varieties of Carpathian walnuts.

Comstock, Ferre & Co.
Box 125T
Wetherfield, CT 06109

This is a very old company and offers some heirloom varieties of seeds. It also has a good selection of herbs.

The Connor Co.
105 N. 2nd St.
P. O. Box 5340G
Augusta, AR 72006

This catalogue lists strawberries exclusively. It does not offer heirloom varieties.

The Converse Co. Nursery
Amherst, NH 03031

This catalogue lists a number of old varieties of apples.

Country Herbs
Box 357
Stockbridge, MA 01262

This catalogue offers plants rather than seeds, which many gardeners would consider an advantage. The listing is not descriptive.

David Crockett Popcorn Co.
P. O. Box 237
Metamora, OH 43540 $.50 plus SSE

This leaflet describes popcorn culture and offers ten different varieties of seed.

De Giorgi Co.
P. O. Box 413
Council Bluffs, IA 51501 $1

The catalogue includes a great deal of information about culture, and a large number of international seeds that few other American companies offer, are listed.

J. A. Demonchaux Co.
827 N. Kansas
Topeka, KS 66608

This company offers a variety of seeds for fresh garden plants. It also sells expensive gourmet varieties.

Dr. Yu Farm
P. O. Box 290
College Park, MD 20740

A good source for Oriental seeds. Such rarities as bitter melon, edible seed watermelons and miniature corn are offered.

Earl Douglas
Red Creek, NY 13143 $.25 for brochure

They offer an American chestnut supposedly resistant to chestnut blight.

Eastville Plantation
P. O. Box 337, Dept. D
Bogart, GA 30662

This is a small listing of nursery stock, specializing in varieties suitable for Southern growing conditions.

Elong's
Box 626
Stevensville, MI 49127

Endangered Species
12571 Redhill
Tustin, CA 92680

A great many of the plants offered in this catalogue are on endangered-species lists. They include succulents, desert plants, jungle plants, etc.

Epicure Seeds
Box 69
Avon, NY 14414

This catalogue offers a wide variety of European vegetable varieties.

Esey's Seeds, Ltd.
Dept. 0G7
York, Prince Edward Island
Canada, COA 1PO

Evans Plant Co.
Box 1649
Gleason, TN 38229

Exotic Seed Co.
1742 Laurel Canyon Blvd.
Hollywood, CA 90046

They specialize in what might be termed tropical fruits, but include pawpaws and persimmons.

Farmer Seed and Nursery
Faribault, MN 55021

A fairly conventional catalogue, although it offers one or two unusual—heirloom or gourmet—varieties for many seeds.

Henry Field Seed & Nursery Co.
Shenandoah, IA 51602
This catalogue offers a wide number of different varieties of most species, including some that are rare or heirloom. It has an unusually complete selection of hybrid sweet corns, as well as some open-pollinated varieties. It also is a source of seed for broom corn. The selection of fruit trees is large, if not unusual. It offers some native fruits, such as pawpaw, serviceberry and persimmon.

Flint Ridge Herb Farm
Rte. 1, Box 187
Sister Bay, WI 54234

Florida's Vineyard Nursery
P. O. Box 300
Orange Lake, FL 32681
They have a large collection of southeastern grape varieties, including the scuppernong grape of southern wine fame.

Fred's Plant Farm
Rte. 1
Dresden, TN 38225
This leaflet offers a very large selection of different varieties of sweet potatoes. It also offers a large selection of chewing and smoking tobacco in bulk quantities.

Garden of Eden Nursery
Rte. 2, Box 1086
Spruce Pine, NC 28777
This nursery specializes in antique fruit trees.

Louis Gerardi Nursery
R. R. 1, Box 146
O'Fallon, IL 62269
They offer a large selection of nuts, and such less-common fruits as pawpaws and mulberries.

Le Marché Seeds International
P. O. Box 190
Dixon, CA 95520 Price: $2.00

As the title suggests, this catalogue is a source for international varieties of seeds. The selection is excellent, and many of the seeds are hard to locate elsewhere, or are completely unavailable. Catalogue entries are clear and informative, and the format is attractive. The seeds tend to be expensive and the quantities small.

Gourmet Gardens
Dept. OGI
923 N. Ivy St.
Arlington, VA 22201

A small but excellent herb catalogue. It includes a great deal of readable information about the cultivation and uses of herbs. Also it has interesting recipes.

Green Tech Nursery
53 Four Mile
Traverse City, MI 49684 $2

The catalogue is a small listing of rare nut trees, specializing in hardy northern varieties. It is one of the rare sources of the American chestnut.

Gurney's
Yankton, SD 57079

This is a large and splashy catalogue which lists a number of unusual varieties of plants. However, there is a great deal of hype in their advertising. For shipping, bare-root trees are not protected at all; I ordered from them a pawpaw and a mulberry that were DOA.

Dean Foster Nurseries
Rte. 2, Dept. OG-1c
Hartford, MI 49057

This nursery list specializes in berries. The varieties of raspberries, red and black, is unusually large. They also list a number of berries not commonly seen, such as Juneberry and Youngberry, and a large listing of currants.

Grace's Gardens
10 Bay St.
Westport, CT 06880 $.50

Joseph Harris Co., Inc.
26 Moreton Farm
Rochester, NY 14624

This is an attractive catalogue listing fairly conventional varieties.

Hastings Southern Garden Guide
Dept. C12, Box 4274
Atlanta, GA 30302

This is an especially good catalogue for southern varieties, both of seeds and of nursery stock. It includes a large number of varieties of pecans and southern varieties of grapes. It has a large number of berries. The catalogue is quite attractive, and there is much information about cultivation and about pollinization requirements where appropriate, and some information about using the produce.

The Herb Shop
Organic Specialty Plants
1942½ Ceffillo's Rd.
Santa Fe, NM 87501

Herbst Seeds, Inc.
N. Main St. 6053
Brewster, NY 10509

This is a large commercial seed company, matter-of-fact and businesslike. It combines attractive color photographs with extensive and very useful charts which list the varieties with their size, days to maturity, and whatever else is pertinent.

Hickory Hollow
Rte. 1, Box 52
Peterstown, WV 24963

Their leaflet lists a small number of herb seeds as well as a variety of dried or otherwise prepared herbs, and herb mixtures for various purposes, including smoking.

Hillier and Sons
Winchester, England

This company has supposedly the largest single selection of old varieties of plants in the world. It does ship to North America. I have not, however, received its catalogue.

J. C. Hollar Seed Co.
P. O. Box 1720
Yuba City, CA 95992

Horticultural Enterprises
P. O. Box 34082
Dallas, TX 75234

This is a company which specializes exclusively in peppers, and its catalogue lists some thirty varieties.

House of Wesley
Nursery Division
Bloomington, IL 67170

A moderate-sized catalogue which offers quite a number of novelty plant and vegetable items and which tends to trade on novelty value.

J. L. Hudson, Seedsman
P. O. Box 1058
Redwood City, CA 94064 $1

A plain catalogue with black-and-white botanical illustrations of great completeness for someone interested in unusual seeds. It includes a large number of Mexican and pre-Columbian varieties, vegetables and herbs. The succinct entries give much information on cultivation, use and matters of general information. It is much respected among those interested in heirloom varieties.

Huvrov's Exotic Seeds
P. O. Box 1596
Chula Vista, CA 92012

Many of the offerings are of unusual houseplants and truly tropical plants. However, some, such as amaranth, are adaptable to temperate areas.

Illinois Foundation Seeds
Box 722
Champaign, IL 61820

This is a small leaflet which lists only hybrid sweet corn.

Internode Seed Co.
P. O. Box 2011
South San Francisco, CA 94040 $1

Jackson & Perkins Co.
Medford, OR 97501

Although most widely known for their roses, they also have listings of various fruits and berries. Varieties are mostly conventional and commercial. Service is exceptionally personal; when they say delivery will be at the proper time for your area, they mean it. Do business with them once and they'll remember you for years. Packaging of live plants is excellent. Nothing I have ordered has ever failed to live.

James Pecan Farm, Inc.
Rte. 3, Box 212
Brunswick, MO 65236

Besides many varieties of pecans, this company offers the hican—a cross which occurred naturally between the pecan and the hickory.

Johnny's Selected Seeds
P. O. Box 202
Albion, ME 04910

This is one of the commercial houses most highly respected by the seed-saving fraternity. It has a small catalogue with an extraordinary selection of heirloom and offbeat varieties. Descriptions are clear and informative. Explanations are superb—better than many gardening books. Service is personal and quick.

Johnson Nursery
Ellijay, GA 30540

This catalogue lists a decently large number of fruit trees, including a number of old varieties of apples, peaches and plums. They claim to grow over one hundred varieties of peaches, so perhaps on special order they may find rare or hard-to-locate varieties. They also sell rootstock and some grafting supplies.

J. W. Jung Seed Co.
Box D-364
Randolph, WI 53956

Although this is a general catalogue, it offers several offbeat old varieties of apples.

Kelly Bros. Nursery, Inc.
316 Maple St.
Dansville, NY 14437

Kelly Brothers Nursery
1408 Sunset Drive
Vista, CA 92083

Kitazawa Seed Co.
356 W. Taylor St.
San Jose, CA 95110

This company offers a small listing, exclusively of Oriental vegetables.

Wm. Knrohme Plant Farms
Rte. 4, Box OG-1c
Dowagiac, MI 49047

This company specializes in strawberries and seems to cater to commercial purchasers, with discount prices for large quantities.

Krider Nurseries
Middlebury, IN 46540

Lawson's Nursery
Ball Ground, GA 30107

They offer a nice selection, although not exhaustive, of old-time fruits, especially apples. Trees I ordered from them for the spring of 1982 were beautiful specimens, remarkably well packed, and grew strongly and well.

Lawyer Nursery
Rte. 2, Box 95
Plains, MT 59859

Basically this is a wholesale nursery, but valuable to those who order in quantity. They sell rootstock, and someone who intends to graft large numbers of trees might find their prices attractive.

Le Jardin du Gourmet
West Danville, VT 05873

A catalogue of a wide variety of gourmet and unusual seeds and bulbs and international varieties. Many hard-to-find varieties are listed. It also offers gourmet foods, such as French shallots, snails, truffles, etc.

Lee-Land Nursery
Box 223
North Kingsville, OH 44068 $1

At one dollar, their booklet, described as offering grafting information and a price list, seems overpriced. They sell grafting materials and a wide selection of rootstock.

Orol Ledden & Sons
P. O. Box 7, Dept. D
Center St.
Sewell, NJ 08080

This is a no-nonsense sort of catalogue, with crisp descriptions and few illustrations. They offer a wider-than-average selection of corn. Their prices seem especially reasonable. They have no nursery stock.

Letherman's, Inc.
1203 Tuscarawas St.
Canton, OH 44707

This is an attractive catalogue with exceptionally large listings of varieties of seeds. It includes few if any heirloom varieties, however.

Henry Leuthardt
P. O. Box 666
East Moriches, NY 11940

This is a small catalogue, but it lists a number of heirloom varieties of apples and other fruits and offers a variety of rootstocks. Service is smooth and personal. Trees I ordered were of first-class quality, and I have never seen better packaging of bare-root trees.

Liberty Seed Co.
P. O. Box 806-B
New Philadelphia, OH 44661 $1

A commercial catalogue with a large listing of fairly usual varieties, but with a minimum of hype. Brief but clear descriptions of varieties and culture make it a useful supplement to gardening books.

Linda Vista Grapevine
Linda Vista Ave.
Napa, CA 94558

This is a very complete catalogue of wine grapes and table grapes.

Makielski Berry Farms
7130 Platt Road
Ypsilanti, MI 48197

This small catalogue lists only such small fruits as raspberries, blackberries, currants, gooseberries, etc.

Margrave Plant Co.
Gleason, TN 38229

This company specializes in sweet potatoes, and it offers about ten varieties.

Earl May Seed and Nursery Co.
2052 Elm St.
Shenandoah, IA 51603

This is a large commercial offering but tastefully free of purple prose and splashy pictures. Generally a fairly usual listing. Berries which I ordered arrived in good shape, but a bare-root nut tree was a poor specimen and very badly packed. I didn't expect it to grow, and it didn't.

Mayo Nurseries
Lyons, NY 14489

They stock a wide selection of most conventional varieties of fruit.

Mellinger's
2380J South Range Rd.
North Lima, OH 44452

Meyer Seed Co.
600 S. Caroline St.
Baltimore, MD 21231

The catalogue offerings are conventional, but they have good information on cultivation in the Maryland area.

Miller Nurseries
321 West Lake Rd.
Canandaigua, NY 14424

This company is reputed to have a number of antique apples, but I did not receive its catalogue.

Nationwide Seed & Supply
Box 91073
Louisville, KY 40291

Naturalist Seed Co.
P. O. Box 435-G2
Yorktown Heights, NY 10598 $1

This is a small company, and its catalogue is particularly informative about the history, use and cultivation of herbs. It is not an exhaustive listing, but most gardeners will find it sufficient.

Neil's Open Pollinated
R. R. 2
Mt. Vernon, IA 52314

This company offers open-pollinated heirloom varieties that the Neil family has propagated for about seventy-five years.

Neosho Nurseries
900 No. College
Neosho, MO 64850

Nichols Garden Nursery
Pacific North
Albany, OR 97321

This is an extremely valuable catalogue for those interested in rare and unusual varieties. It has heirloom varieties as well as gourmet varieties from all over the world. It also includes a great deal of useful information about cultivation and consumption. I have found their service quick and have had good luck with their seeds.

Nourse Farms, Inc.
Box 485, Dept. OG12
South Deerfield, MA 01343

This nursery specializes in berries. The listing is fairly large, but conventional. It has a good section on berry cultivation.

Olds Seed Co.
Box 7790, Dept. N
Madison, WI 53707

This seed listing is fairly conventional, although they list a few unusual seeds, such as sugar beets and a larger variety of potatoes than is usually found commercially.

Oregon Bulb Farms
39391 S.E. Lusted Road
Sandy, OR 97055

This is a wholesale catalogue, with a minimum order of $1,500. Besides bulbs they offer a very wide variety of fruits and ornamentals, although mostly of usual varieties.

Ozark National Seed Order
Drury, MO 65638 SSE

This company offers open-pollinated varieties of plants, and its seed is untreated.

Geo. W. Park Seed Co., Inc.
86 Cokesbury Rd.
Greenwood, SC 29647

This is a respectable catalogue of standard varieties of vegetables. It is an attractive catalogue, and has a particularly good selection of flower seeds.

Pellett Gardens
Atlantic, IA 50022

This is a catalogue of plants and shrubs to which bees are particularly attracted.

Piedmont Plant Co., Inc.
Dept. 103, Box 424
Albany, GA 31703

Pike's Peak Nurseries
R.D. 1
Penn Runn, PA 15761

This nursery offers a limited selection of fruit trees and has a good selection of landscaping trees, particularly evergreens.

Pinetree Seed Co.
P. O. Box 1399
Portland, ME 04104

This small catalogue lists a fairly conventional line of seeds. They specialize in small packets of seeds, which are priced appropriately below full-sized seed packets. A small gardener who never uses a full packet might consider this a good idea.

Pony Creek Nursery
Tilleda, WI 54978

This tabloid-sized catalogue on newsprint offers a few heirloom apple trees, but for the most part the plant and seed varieties are usual. Seed prices are quite low. The catalogue also contains chatty and useful information and lore.

Ponzier Nursery
Rte. 1, Box 313-C
Lecoma, MO 65540

Ray and Peg Prag
Forest Farm
990 Tetherow Rd.
Williams, OR 97544 $1

This company offers a wide variety of plants, including wild plants and many unusual plants.

Ben Quisenberry
Syracuse, OH 45779

Send a self-addressed envelope for this seed list. He is a person, not a company, and offers a few varieties of tomatoes, all open-pollinated and mostly heirloom or unique. The service is impressively personal. He is respected and revered by everyone in the seed-saving group.

Raintree Nursery
265 Butts Rd.
Morton, WA 98356

This is a small catalogue, but it offers a number of interesting varieties of apples. It also offers a number of hard-to-find Oriental pears and has a special listing for wine grapes.

Rayner Bros., Inc.
Dept. 98
Salisbury, MN 21801

Redwood City Seed Co.
P. O. Box 361
Redwood City, CA 94064 $.50

This company offers a wide variety of all kinds of garden plants and trees, including foreign varieties. It has many hard-to-find heirloom plants, and the catalogue provides a historical perspective of the origins of many plants.

Otto Richter and Sons
Box 26
Goodwood, Ontario LOC IAO
Canada

This is one of the preeminent herb companies in the world, listing over three hundred varieties.

Judd Ringer
6860 Flying Cloud Drive
Eden Prairie, MN 55344

Roses of Yesterday and Today
802 Brown's Valley Rd.
Watsonville, CA 95076 $1.50

This is an elegant but not splashy catalogue with a very large listing of roses, each with the date of introduction or of first description. It also lists societies and organizations interested in old varieties, as well as books and periodicals. Includes much information about rose culture and cultivation. The text is literate and tasteful.

St. Lawrence Nurseries
Canandaigua, NY 14424

This company is reputed to stock many antique fruits, but I did not receive a catalogue.

Sanctuary Seeds
2386 W. 4th
Vancouver, BC
Canada V6K 1P1

They offer open-pollinated seeds and a selection of herbs. They also engage in seed exchange.

Sandy Mush Herb Nursery
Rte. 2, Surrett Cove Rd.
Leicester, NC 28748 $1 (refundable)

This is an attractive catalogue offering both herb seeds and plants. It includes recipes, growing instructions and interesting herb-garden plans and layouts.

Savage Farms Nursery
P. O. Box 125
McMinville, TN 37110

This catalogue offers a number of varieties especially suitable for Southern cultivation, including varieties of fruits, such nuts as hickory, pecan, hazelnuts, scuppernong grapes, etc.

Seedway, Inc.
Box 250
Hall, NY 14463

As the name indicates, this company offers only seed. The varieties are conventional but extensive.

Self-Sufficient Seeds
Barr Hollow
Woodward, PA 16882 $2

A good source for traditional varieties of grains, including corn and beans, as well as rare small grains. The text of the catalogue is exceptionally informative, both about varieties and about their cultivation.

R. H. Shumway Seedsman, Inc.
Attn: Mrs. Gillette
628 Cedar St., P. O. Box 777
Rockford, IL 61101

This is a very old company and offers a number of sometimes difficult-to-locate varieties. It is one of the few catalogues which offers sorghum, and more than one variety. They also have a good selection of grapes.

Southern Garden Co.
P. O. Box 745
Norcross, GA 30091

This is a plain, no-nonsense catalogue which specializes in varieties especially popular in, or suitable to, the southern part of the country. It includes, for example, varieties of cowpeas seldom found in other catalogues, southern grapes, and varieties of fruit trees suitable for growing in areas with minimum winter chill.

Southmeadow Fruit Gardens
2363 Tilbury Place
Birmingham, MI 48009 $8 for their directory (and well worth it); price list free.

An excellent collection of rarities and an unusually full and complete description of each variety, together with copious quotations from ancient, old and not-so-old commentators. The selection is superb, but the prices of stock are quite high.

Spring Hill Nurseries
Catalogue Reservation Center
P. O. Box 1758
Peoria, IL 61656 $1

This is a attractive catalogue. It specializes in flowering plants, of which it has many unusual varieties. There is a small listing of fruits, nuts and berries, and a succinct listing of herbs.

Stark Bros. Nurseries and Orchards Co.
Box J3421B
Louisiana, MO 63353

This nursery has fruit varieties especially suitable for southern planting. The listing is otherwise fairly conventional. The catalogue is especially informative.

Steele Plant Co.
Gleason, TN 38229

This company specializes in sweet potatoes and has a large listing of varieties.

Stern's Nurseries, Inc.
Geneva, NY 14456

Their listing is small and assorted. Prices seem a bit high.

Stokes Seeds
1142 Stokes Bldg.
St. Catharines, Ontario L2R 6R6
Canada

This highly respected Canadian company offers a large variety of everything. Although it does not especially feature heirloom varieties, it has many varieties most American gardeners might consider unusual or gourmet.

Sunrise Enterprises
P. O. Box 10058
Elmwood, CT 06110

A small catalogue listing exclusively Oriental vegetables. Besides giant radishes, green onions, etc., it includes such exotica as Chinese edible seed watermelons, bitter melons and minature corn.

Sunsweet Berry and Fruit Nursery
Sumner, GA 31789

Sutter's Apple Nursery
3220 Silverado Trail
St. Helena, CA 94574

They have some seventy varieties of antique apples. Send SSE for their list.

Talbott Nursery
R. R. 3, Box 212
Canby, OR 97013

Taylor's Garden
1535 Lone Oak Rd.
Vista, CA 92083

They offer about two hundred varieties of herbs, sold as plants rather than seeds. Since many varieties of herb seeds are difficult to germinate, this is a valuable source to know.

Thompson and Morgan
Dept. OG3/82
P. O. Box 100
Farmingdale, NJ 07727

This is the most beautiful catalogue I've ever seen. The photographs are absolutely stunning. They have an especially large and varied selection of flowers. The selection of vegetables is good, but generally conventional. The catalogue has good brief instructions for cultivation, plus scattered tips on the use of various crops.

Tree Crop Nursery
76500 Short Creek Rd., No. 4
Covelo, CA 95428

This nursery offers a listing of varieties of pears, apples and plums. The number of varieties is not large, but it includes rarities of each species, including greengage plums and several uncommon old apples.

Otis S. Twilley Seed Co.
P.O. Box 65
Trevoose, PA 19047

Urban Farmers, Inc.
22000 West Haliburton Rd.
Beachwood, OH 44122

This small catalogue specializes in European and Oriental varieties, and varieties that I have seen nowhere else are listed. The catalogue is also informative and provides special information about cultivation of the individual varieties. I have been well satisfied with their seeds, but the size of the packets and the quantity of seeds is niggardly. One who orders lettuce or carrots, for example, might do well to order three or four packets rather than one.

Van Bourgondien Bros.
Box A-OG1
Rte. 109
Babylon, NY 11702

A very attractive catalogue with a good selection of bulbs. It is one of the few catalogues I have seen which offers saffron bulbs from which the rare condiment saffron is made.

Vermont Bean Seed Co.
Garden Lane
Bomoseen, VT 95732

As the title might suggest, this is a no-nonsense sort of catalogue. It lists all the common types of vegetables, with no particular emphasis on heirloom varieties. However, it has a listing of traditional bean varieties that is most ample, including descriptions of the bean and its cultivation as well as of its culinary attractions.

Waynesboro Nurseries
Waynesboro, VA 22980

They carry a number of old varieties of apples and supply the extremely rare American chestnut. They also offer chinquapins.

Wayside Gardens Co.
127 Garden Lane
Hodges, SC 29695

Well-sweep Herb Farm
317 Mt. Bethel Rd.
Port Murray, NJ 97865

Dave Wilson Nurseries
Box IOG
Hughson, CA 95326

Wolf River Nurseries
Rte. 67, Box 73
Buskirk, NY 12028

This catalogue offers a small listing of a select number of trees and berries, specializing in hardy varieties. It offers a hybrid chestnut which is a cross between the Chinese chestnut and the American.

Wyatt-Quarles Seed Co.
Box 2131
Raleigh, NC 27602

This catalogue offers an unusual number of different varieties, especially, for example, field peas, okra and green beans.

Yankee Peddler Herb and Health Farm
Rte. 1, Box 251A, OG
Burton, TX 77835 $1

This catalogue specializes in herbs and health plants as well as health products. Besides herbs it lists wild strawberry seeds, saffron bulbs and similarly offbeat and difficult-to-locate varieties. It also has extensive offerings of bulk herbs and a considerable listing of books.

APPENDIX II
DIRECTORY OF FOUNDATIONS

The following directory of foundations, organizations and periodicals interested in various aspects of heirloom gardening and preservation of endangered species is not at all complete. It does, however, include what I consider the major noncommercial sources of rare and unusual varieties of fruits and vegetables. Most of them are primary sources. Although some, such as the linchpin Seed Savers' Exchange, are a direct source of plant materials for members, many, including the SSE, provide contacts with individuals of like interests, from whom seeds and scion may be located on a barter basis. Membership in these organizations is usually very inexpensive, and for anyone whose interests incline them at all in those directions it is certainly an interesting investment.

Abundant Life Seed Foundation
P. O. Box 772
Port Townsend, WA 98368 $2

This group publishes a catalogue of vegetables, etc., and specializes in Pacific Coast plants. The $2 fee includes a subscription to their quarterly newsletter.

American Fruit Grower
Meister Publishing Co.
37841 Euclid Ave.
Willoughby, OH 44094 $10 per annum

American Gourd Society
P. O. Box 274
Mount Gilead, OH 43338 $2.50

The membership fee includes a year's subscription to *The Gourd,* featuring a seed exchange of rare and noncommercial varieties.

American Horticulturist
Yankee Peddler Herb and Health Farm
Rte. 1, Box 251A
Burton, TX 77855 $20

The membership fee includes six issues of their publication.

American Pomological Society
103 Tyson Bldg.
University Park, PA 16802

Butterbrooke Farm
Oxford, CT 06483

This is a co-op which offers seeds only to members of the exchange. They have many heirloom varieties, and all are open-pollinated.

California Rare Fruit Growers
Star Route Box P
Bon Sall, CA 92003

New and Noteworthy Fruits
New York State Fruit Testing Cooperative Association
Geneva, NY 14456

A catalogue of hundreds of fruits and berries of all kinds.

Family Food Garden (1 year, 6 issues)
Yankee Peddler Herb and Health Farm
Rte. 1, Box 251A
Burton, TX 77835

The Graham Center
Seed Directory
Rte. 3, Box 95
Wadesboro, NC 28170 $1

This is a nonprofit organization operated by the Rural Advancement Fund of the National Sharecroppers Fund. It provides an annotated list of seeds, as well as a bibliography. It also gives brief but good information about seed saving, etc. It is a very important and useful guide and certainly deserves the support of all who are interested in endangered species.

Home Orchard Society
2511 S.W. Miles St.
Portland, OR 97219

International Association for Lesser Known Food Plants and Trees
P. O. Box 599
Lynwood, CA 90262

International Dwarf Fruit Tree Association
Horticulture Department, Michigan State University
East Lansing, MI 48824

Island Foundation Tree Crop Nursery
Rte. 1, Box 44B
Covelo, CA 95428

National Arbor Day Foundation
Arbor Lodge 100
Nebraska City, NE 68410

North American Fruit Explorers
Mr. Michael McConken, Secretary
2519 Coolspring Rd.
Adelphi, MD 20783 $5

They have an annual publication: *North American Pomona.* For those interested in old varieties of fruits and nuts, this is the indispensable organization to join. Membership includes a subscription to *The Pomona,* a superb magazine of information and exchanges of information, organizations and plants.

Northern Nut Growers' Association
Attn.: John H. Gordon, Jr.
1385 Campbell Blvd.
North Tonawanda, NY 14120
They promote the collection of pecans hardy in the North.

Plant Finders of America
Dept. 10A
106 Fayette Circle
Fort Wright, KY 41011

Rare Fruit Council
Museum of Science
S. Miami Ave.
Miami, FL 33129

Seed Savers' Exchange
R. R. 2
Princeton, MO 64673 $3
This is a nonprofit organization and is the preeminent organization of the guerrilla gardeners. It publishes a yearbook containing articles and a list of members with the seeds they have to offer and the varieties they want. Even people not especially interested in seed exchanging or growing heirloom varieties would find it worth joining.

Soy Craft Magazine
158 Main St. No. 3
Greenfield, MA 01301 $15
One year's subscription includes four issues. This magazine is dedicated to the growing and using of soybeans.

Stronghold, Inc.
4931 Upton St. N.W.
Washington, DC 20016 $10
This is a nonprofit organization concerned with preserving the American chestnut.

U.S. Department of Agriculture
Northwestern Region
Dr. M. Faust, Fruit Lab PGGI
Beltsville, MD 20705

This office maintains a list of most varieties of old-time fruits and nuts grown in the United States.

Wanigan Associates
262 Salem St.
Lynnfield, MA 01940 $5

This organization catalogues over four hundred bean varieties, with descriptions, from the largest private bean collection ever, that of the near-legendary John Withee. The fee includes the catalogue and a subscription to *The Wanigan—A Bean Newsletter.*

Worcester County Horticultural Society
30 Elm St.
Worcester, MA 01608

This society maintains an antique-apple orchard and sells scion. It does not sell trees, but will send a free list of varieties if a self-addressed stamped envelope is enclosed.

APPENDIX III BOOKS RELATED TO HEIRLOOM-GARDENING INTERESTS

Any bibliography that would attempt to account for all books and all varieties of books and publications that would appeal to a guerrilla gardener would be impossibly complex and convoluted, and the bibliographer who compiled it would end up a raving lunatic. He would have been a lunatic to begin it in the first place. There is no end to the variety of works that would appeal to the guerrilla gardener, who is by his very nature and definition limitless in his interests. The only common denominator is a passion to *grow* things (not grow *things*).

The following list of books is a heterogeneous one selected only on the basis of that common denominator. It includes further information and suggests areas of study of all of the myriad and disjointed topics touched on in this book. It's quirky because the author of this book is quirky. The pamphlets published by commercial houses which promote their own products as well as provide information should not be scorned; their information is directed toward the beginner in most cases, and explanations begin from the bottom and go up. If a person wishes to own one, most of them are free. Many publications listed are from state universities or agricultural research stations. There are many more such publications than are listed. There are also books and registers published and kept current by research centers of the agricultural colleges. I have not attempted either completeness or balance in these listings; I list pomological registers, for example, but not registers of corn, on the basis that involvement with the myriad varieties of corn is less a possible individual pleasure than it is with fruit. I hope this does not offend corn fanciers; growing more than a few varieties of corn begins to become research, whereas a sizable fruit orchard can still be maintained essentially for the fun of it. Something of the same rationale applies to other inclusions and exclusions.

Most but not all of the publications listed I have examined. Many are referred to in the text of this book. Some not examined were included because their titles indicated that someone interested in guerrilla gardening would find them useful; sorghum, for example. I have made annotations to many which I have examined; others I have let stand by themselves. Like discovering exotic grafts and grafting techniques, many pleasures of bibliographic discoveries and explorations are keenest if left to the interested individual.

All About Growing Fruits and Berries. Ortho Books.

All About Pruning. Ortho Books.

American Pomological Registry. U.S. Department of Agriculture, Washington, D.C. This is the official USDA description of apples. It is safe to call it exhaustive.

Beginning a Backyard Orchard. West Virginia Department of Agriculture, Charleston, W.V.

Boswell, U. R., and L. B. Reed, *Okra Culture.* U.S. Department of Agriculture leaflet, No. 449 (1962).

Bright, Lloyd, *Producing Giant Watermelons.* This small booklet, published by a grower of record watermelons, explains techniques for growing truly huge watermelons and includes considerable watermelon lore.

Burpee's 1888 Catalogue. W. Atlee Burpee Co., Warminster, Pa. This reproduction of the early Burpee catalogue is not only attractive, it is also useful in helping identify and date varieties.

Burpee's Nursery Guide. W. Atlee Burpee Co., Warminster, Pa.

Correll, D. S., *The Potato and its Wild Relatives.* Renner, Texas, 1962.

Darrow, G. M., *The Strawberry: History, Breeding and Physiology.* New York, 1966.

Disease and Insect Control in the Home Orchard. Cornell University, Ithaca, N.Y.

Doggett, H., *Sorgum.* London, 1970.

Frankel, O. H., and E. Bennett, eds., *Genetic Resources in Plants—Their Exploration and Exploitation and Conservation* Philadelphia, 1970.

Fruit and Tree Nut Germ Plasm Resources Inventory. Information Office, Agricultural Service, U.S. Department of Agriculture, Washington, D.C. 20250. Lists all fruit and nut varieties now growing on government research stations.

Future Is Abundant, The. Arlington, WA: Tilth, 1983. This paperback has numerous articles on the general topics of guerrilla gardening, and a fine listing of catalogues, many off-beat, of interest to guerrilla gardeners.

Garner, R. J. *The Grafter's Handbook,* 2nd ed. New York: Oxford University Press, 1979. This is the standard book on grafting. It includes a bibliography of 135 items related to grafting.

Genetic Vulnerability of Major Crops. Washington: National Academy of Sciences, 1972.

Gregory, W. C., *et. al., Peanuts—Culture and Use.* Stillwater, Okla., 1973.

Hartman, H. T. and D. E. Kesler, *Plant Propagation.* Englewood Cliffs, N. J.: Prentice-Hall, 1960. This volume includes information on starting plants from cuttings, layering, starting difficult-to-germinate seeds, etc.

Heiser, C. B., *Nightshades, the Paradoxical Plants.* San Francisco, 1969. This review of the entire family includes information about tomatoes and potatoes.

———, *The Sunflower.* Norman, Okla.: University of Oklahoma Press, 1975. This is considered the authoritative book on the subject of sunflowers.

Hendrick, Hall, Hawthorn and Berger, eds., *Vegetables of New*

York. New York, 1928. Because of its age, this is an important reference for identifying old varieties of vegetables.

Herklots, G. A. C., *Vegetables in Southeast Asia.* London, 1972. This work is useful in tracing vegetables of Oriental origin.

The Home Fruit Planting. Cornell University, Ithaca, N.Y.

Home Vegetable Garden Pest Control. West Virginia Department of Agriculture, Charleston, W.V.

Hutchinson, J. B., ed., *Essays on Crop Plant Evolution;* Cambridge, Mass., 1965.

Insect and Disease Control on Vegetables. Cornell University, Ithaca, N. Y.

Insects and Diseases in the Home Vegetable Garden. Cornell University, Ithaca, N. Y.

Jaynes, R. A., ed., *Handbook of North American Nut Trees.* Geneva, N. Y., 1969.

John Bacon Corp. publication, Gasport, N. Y. 14067. This publication offers information about grafting and budding and is also a catalogue of grafting supplies.

Johnston, Robert, Jr., *Growing Garden Seed.* Albion, Maine: Johnny's Selected Seeds, 1976. This booklet, published by the founder of *Johnny's Select Seeds,* gives a brief, brisk account of the theory and practice of seed saving.

Kline, Roger A., Robert F. Becker and Lynne Belluscio, *The Heirloom Vegetable Garden.* Cornell University, Ithaca, N. Y. This attractive booklet lists descriptions and reproduces illustrations from old catalogues.

Leonard, A. M., and Sons, Inc., publication, Piqua, Ohio 45356. Their booklet provides information on budding and grafting and includes a catalogue for budding and grafting supplies.

Lipnack, Jessica, and Jeffrey Stamps, *Networking: The First Report and Directory.* Garden City, N. Y.: Doubleday, 1982. A listing of everything from directories to bed-and-breakfast places to a society for agoraphobics. It includes organizations involved with propagation of rare seeds, and some catalogues. It is pretentious and also difficult to use.

Mangelsdorf, P. C., *Corn: Its Origin, Evolution and Improvement.* Cambridge, Mass., 1974.

Mellinger's Budding and Grafting Supplies, published by Mellinger's, North Lima, Ohio. This publication includes information about budding and grafting and includes a catalogue of budding and grafting supplies.

Nitschke, Robert A., *Southmeadow Fruit Gardens, Choice and Unusual Fruit Varieties.* Southmeadow Fruit Gardens, Birmingham, Mich. This is something of a hybrid between a reference work and a catalogue. The reference work explains with amazing detail and completeness the appearance and origins of many old varieties of fruits. It has to be something of a bible to growers of heirloom fruits. Accompanying it is a price list, from which all listed varieties may be ordered. They tend to be a little pricey.

North, C., *Plant Breeding and Genetics in Horticulture.* Washington: U.S. Department of Agriculture, n.d. This government publication gives an introduction to practices some guerrilla gardeners may wish to master and is not exhaustively technical.

Peach Growing. Cornell University, Ithaca, N.Y.

Poisonous Plants. West Virginia Department of Agriculture, Charleston, W. V.

Plants for Difficult Situations. University of Connecticut, Storrs, Conn.

Plants with Poisonous Properties. University of Connecticut, Storrs, Conn.

Rehder, A., *Manual of Cultivated Trees and Shrubs,* 2nd ed. New York, n.d.

Rogers, Mark, *Growing and Saving Vegetable Seeds.* Charlotte, Vt. n.d. Besides the mechanics of growing seeds, etc., this includes instruction for hand pollinization.

Renfrew, J. M., *Palaeoethnobotony—The Prehistoric Food Plants of the Near East and Europe.* London, 1965.

Royal Horticultural Society, *The Fruit Garden Displayed,* rev. ed.

London, 1965. While it is concerned with British varieties, much of the information is general. It is especially useful for information about the pollinization requirements of different varieties of fruit.

Simmonds, N. W., ed., *Evolution of Crop Plants.* London and New York: Longman, 1976. This book contains essays on virtually all crop plants, each written by an international expert. Each essay includes technical genetic information; although their literacy is not uniform, all of the essays are readable. It also includes a bibliography for each essay; much of the material is technical, but much isn't. If this book is not indispensable, it is certainly irresistible.

Streuver, S., ed., *Prehistoric Agriculture.* New York: American Museum of Natural History, 1971.

Tomkins, J., and G. Aberly, *The Home Fruit Planting.* Cornell University, Ithaca, N.Y. This is a general fruit-growing book. It includes information about rootstock, pollinization, etc.

Tukey, H. B., *Dwarfed Fruit Trees.* Ithaca, N. Y.: Cornell University Press, 1964.

Ucko, P. J., and G. W. Dimbley, eds., *The Domestication and Exploitation of Plants and Animals.* London, 1969.

Walker, Ray K., *Nut Growing in Illinois.* This booklet includes information about seed care, grafting, varieties of walnuts, pawpaws, etc.

Wallace, H. A., and W. L. Brown, *Corn and Its Early Fathers.* East Lansing, Mich., 1956.

Weatherwak, P., *Indian Corn in Old America.* New York, 1954. This very readable book describes both the cultivation and the importance of corn in pre-Columbian America. It also explains the genetic mysteries in the ancestry of corn.

Wilhelm, S., and J. Sagen, *History of the Strawberry from Ancient Gardens to Modern Markets,* 1975.

INDEX